HOW TO BUILD A PROJECT
Things They Don't Teach

DEAN:
Thanks for all your help and good luck with your operation

Jeff Bassett

Copyright © 2015 Mr. Ted H. Bassett, Jim Cooke, and Jennifer Parks

All rights reserved, including the right to reproduce
this book or portions thereof in any form whatsoever.

Designed by Jennifer Roberts

Published in Canada.
Printed by CreateSpace, An Amazon.com Company
Printed in the United States of America

1 2 3 4 5 6 7 8 9 10

ISBN: 1505380847
ISBN 13: 9781505380842

HOW TO BUILD A PROJECT
Things They Don't Teach

TED BASSETT, JIM COOKE, AND JENNIFER PARKS

DISCLAIMER

Characters and events in this book are all based on history. However, to bring a scene or person to life, some dialogue was made up, care being taken to offer a realistic representation of people and events. They are opinion.

In some cases, Jennifer Parks created scenes between Jim and Ted from her imagination, based on her interview time with them, in an effort to link the stories and lessons into a coherent narrative.

All characterizations of people and event details derive from the memories and perspectives of Ted Bassett and Jim Cooke. All facts around major political and business events were verified by public record during the writing process. Some names of individuals have been changed out of respect for their privacy.

ACKNOWLEDGMENTS

This book would not exist without the incredible support and encouragement of many people. Thank you, first of all, to Anita, for the original sparks of inspiration for penning these stories and adventures. Your undying strength, courage and pure joy for life are the true lights which guided this project. To all of our amazing families, Mona and Paul Bassett; Ashley Fulks and The Parks; Beryl Cooke & clan: Jim, Jenn and I never could have done it without your love, patience and input along the way. We are especially grateful for the Cookes' incredible hospitality during our time together in Kelowna. Thank you to my mentor and friend, John Somers: you set the highest example for the kind of leader I wanted to be. To Gord Clark, thank you for your valuable contributions on the Russian and Canadian mine sampling process. To Jenn Roberts, for the fine cover design and book layout. And to Veronica Rossos, for your professional legal counsel. Finally, thank you to our editors and layout staff at CreateSpace.com for guiding us through the publishing process. It takes a great team to build a successful project. I'm proud of what we've all created together.

- Ted Bassett

CONTENTS

Overview 1

CHAPTER 1
The Promise 5

CHAPTER 2
Commitment 17

CHAPTER 3
Sleeping on the Ore 21

CHAPTER 4
On Being Principled 33

CHAPTER 5
Dodging Bullets 43

CHAPTER 6
Memories 55

CHAPTER 7
Kyrgyzstan Assignment 57

CHAPTER 8
On the Way 61

CHAPTER 9
Master Checklist 65

CHAPTER 10
Not Just a Day 69

CHAPTER 11
Safety and the Law 81

CHAPTER 12
Finding Balance 93

CHAPTER 13
Siberia 101

CONTENTS

CHAPTER 14
Diavik 109

CHAPTER 15
Great Barrier Grief 113

CHAPTER 16
Who's in Charge? 129

CHAPTER 17
Diagnosis 133

CHAPTER 18
Too Many Wrongs 137

CHAPTER 19
La-La-Land 143

CHAPTER 20
Courage 147

CHAPTER 21
The Challenge 155

CHAPTER 22
Ideas 163

CHAPTER 23
Good Project, Bad Timing 171

CHAPTER 24
October 5 177

CHAPTER 25
Friends 179

APPENDIX 1
Project Management Principles 183

Portfolio of Pictures 185

Overview

Ted Bassett has worked his share of projects—from a remote open-pit operation in post-Communist Kyrgyzstan to Busang, Indonesia's scandal-racked Bre-X gold mine. And he's seen it all.

"After witnessing slip-ups that ruined reputations, senseless risk taking that resulted in lost lives and limbs, and project managers who'd fallen in love with their projects, only for their brains to fall out," the project vet wondered, "how the heck did I get here?"

His question was part existential, part tongue-in-cheek. After more than forty years working globally scaled projects in the mining and metals industry, Ted knew well that no sterile textbook full of contrived artificial examples and oversimplified theories could have prepared him for this down-and-dirty cutthroat world. It took some hard-won street smarts, sweat, stamina, and sacrifice to rise to the top of his game—this engineer and married father of two who joked that he couldn't "stand on a ladder and screw in a lightbulb."

But Ted Bassett knew where to find the guy who could, which made him invaluable. He also knew that projects were built by teams, not individuals, and so led his teams accordingly.

Ted credits his successes to the exceptional people he had the good fortune—and good sense—to work with along the way and to his great mentor, John Somers, a realist, who told him: "You might as well aim high, 'cause you're gonna hit low enough, anyway." But even more than his successes, Ted credits his failures for teaching him how to lead and inspire others toward high performance, integrity, loyalty, and teamwork. After decades, all of this got Ted thinking, "Now, how can I give back?"

What advice or guidance could he offer new engineers to help them not only survive but succeed in the muddy snake-filled trenches of project work? Little did Ted realize, the answer to this question lay in his heart.

Years back, his daughter, Anita, had been relentless in urging him to pen his many globe-trotting adventures, which have taken him from soaring above perilous summits to tasting bittersweet triumph to wading in dirt, defeat, and even death. *How to Build a Project: Things They Don't Teach* is that proffered book. Written in a dramatic narrative style, it follows Ted through the peaks and pitfalls of a demanding career in project management, as the concurrent story of Anita's heart-wrenching battle with cancer unfolds.

The reader travels with Ted, family, and friends from the mountains of northern Kyrgyzstan (where Ted's illusions of safety are shattered during a crisis that leaves Anita wondering if she still has a dad), to a tense Saskatchewan court Discovery Room where Ted must face his own demons, on to the frigid North, full of glistening Diavik diamonds, where success and boredom meet, and then Down Under on two of the biggest and most hair-pulling projects of Ted Bassett's epic career.

One day, Ted might be halfway around the world, in a hard hat and work boots, juggling sticky foreign politics with schedule changes, budget constraints, new drill data, and the emergent needs of a three-hundred-person work crew. The next day, he might be power-suited up in a Toronto boardroom for a high-level corporate strategy meeting to solve the mind-bending puzzle of how to save billions during a fiscal crunch while expanding output at one of the planet's largest copper and uranium mines.

Ted's cousin-in-law, Jim Cooke, is another key participant in this book. A fellow engineer who worked in the oil and gas industry, Jim built his career upon vision and leadership. Early on, he'd identified an attitude in the corporate culture that allowed engineers to "sink or swim." Finding this unacceptable, Jim made it his top priority and passion to offer this preparatory training or support. Whereas Ted was hired "to think" and take mining projects from their inception through to their completion, Jim worked with companies and trained their project managers how to think, setting them up for success.

In the summer of 2012, Ted and Jim met for a week in Kelowna, British Columbia, and together, with freelance journalist and creative writer, Jennifer Parks, they fleshed out a concept for this book. After much storytelling and discussion of project principles and philosophies, they came to the conclusion that, despite their different career paths, when you rise above the pieces that go into a project, the principles of sound management are the same, no matter the industry.

This book is the result of that week together.

Says Ted: "Every project starts with a plan and requires blueprints, a budget, a time line, a team, and a leader. Any textbook can teach you that stuff. But only a great manager can give a project vision by inspiring commitment, innovation, leadership, and excellence." These are what you need to draw on, he concludes, when reality inevitably swoops in and hijacks your "perfect" plan.

"I'm hopeful that *How to Build a Project* will give engineers and managers some entertaining late-night reading that still qualifies as work, along with some practical tools and insights for those who are seeking an edge in their profession," says Ted.

Note: At the end of each chapter, you will find a list of principles that apply to the stories and examples found within that chapter and a comprehensive list at the end of the book for easy reference.

CHAPTER ONE

The Promise

"So…what's up, there, Sis?" Paul leaned in, across the dinner table, eyebrow raised as he studied Anita. "You look…somehow…I dunno…different."

"Me? What? No. All the same here." His sister picked up her fork and shoveled in some of her mom's famous cranberry-walnut stuffing. "Mmm…so good."

She chewed deliberately, hand covering her mouth as it twisted up at one corner against her will.

Narrowing his eyes, Paul shook the turkey leg he held in his greasy mitt at her. He tilted his head, staring at his sister more intently. "No, there's something…I'm sure of it. I just can't quite put my finger on it."

Ignoring her brother, Anita whipped her head around toward her mom, utensil wagging in the air. "You outdid yourself, Mom. This is truly amazing."

"Um, thanks, honey…you did help me make it, remember? Early this morning!"

Anita's eyes widened and her cheeks grew pink. "Yeah. Right…that seems like ages ago…Well, we certainly outdid ourselves then, didn't we?"

"Uh-huh." Paul crossed his arms, settling back in his chair. "I rest my case."

Ted and Mona exchanged a puzzled look. Their twenty-six-year-old daughter certainly was behaving strangely. She was typically a happy girl—level headed, busy, and involved, whatever the activity or conversation. There was something different about her. She was flightier than usual, more distracted, apparently harboring some secret delight, ever since she'd arrived home for the holidays. Had something happened over in Prince George at her new accounting job with KPMG that she wasn't telling them?

Before her parents could pry, Anita was off on another tangent.

"Hey, Dad."

"Hey, Anita," Ted teased, taking a sip of Bordeaux. Whatever was up with his daughter, he was enjoying himself immensely. It was so nice to be on holidays with his family again.

"I was thinking..."

"You were thinking?" Paul cut in, razzing his sister.

She playfully took the bait, scowling back, like they were teenagers again.

"Nice hair, skid. Don't they have barbers up in the sticks?"

"It's called lumberjack chic. Don't you like it?" Paul ran a hand across his fuzzy jowl.

Ted winked at Mona, whose face was flushed from the day's activities. He could tell how happy his wife was to have them all back under one roof again. In June, Anita had moved away to take a job in northern British Columbia, more than 1,200 kilometers and two provinces away. Paul was more of a homebody at heart. He'd turned twenty-four that year and had begun attending Forestry College in Prince Albert, a little over an hour's drive away. Keeping with tradition, they'd all packed their bags and made the trek out to the family's rustic cottage at Candle Lake for the holidays.

And that afternoon, honoring another Bassett Christmas family ritual, they'd bundled up and hiked into the bush to cut down a tree. Paul and Ted had hauled it back and set it up, and the girls, mostly, had dressed it, while the boys sipped rum and eggnog, falling into idle chitchat by the fireside while failing miserably at pretending to be useful.

Ted looked around the table and sighed contentedly. Moments like these, together, were now more precious and rare. He had to admit, he'd never seen his daughter quite so blissfully out of sorts. But whatever it was, he wasn't worried. Anita had a good head on her shoulders.

Out the window, a light dusting of snow was falling. It was minus 20 degrees Celsius that night, but they were warm and cozy inside. Ted was grateful for the calm, quiet respite of a few days amid the snow and trees, away from the demands of his work. For the first time in a while, Ted felt optimistic.

"Earth to Dad, come in, Dad!"

Ted smiled at his daughter, raising one eyebrow. "As you were saying..."

"As I was saying...I was thinking...You know what you should do?" Anita leaned in, almost toppling her glass.

Paul chuckled, rubbing his hands together. "This should be good."

Anita shot her brother an annoyed look, and Mona sat back and sipped her wine, thoroughly amused.

"I'm serious, you guys!" Her face was deliberately earnest.

Ted decided to go along with her decoy game. Soon enough they'd get her news out of her.

"OK, what is it that you think I should do, Anita?"

"You should write a book!"

"A book? Why on earth would I do that?" Ted laughed, balking at the notion. "I haven't done anything interesting enough to put in a book."

"Are you kidding me?" Anita's eyes blazed with a challenge. She looked wholeheartedly serious about this idea. "You've been on so many crazy adventures working on your projects all around the world...Kumtor...Bre-X...Saudi Arabia...Russia!"

If this was Anita's attempt to change the subject, she was doing a darned good job. Paul rubbed his chin, thoughtfully.

"I like it. What a novel idea, Sis!" Then he grunted, self-amused. "Pun intended, of course."

Anita groaned, and Mona rolled her eyes.

"Ted Bassett's Adventures in Project Management! It's got a good ring to it. Dad, you should totally do it!"

Picking up his fork, Paul willfully attacked the mountain of mashed potatoes and gravy on his plate. Mona smiled.

"You do have some pretty good stories up your sleeve, hon."

Ted scratched his head. "You think so? I dunno."

The idea of writing a book had never occurred to him, though secretly it put him over the moon that his family thought his experiences worth publishing. For years, he'd regaled them with his far-flung tales around the dinner table and as entertainment on commutes to and from the curling rink. But why bother writing them down? It was as if Anita read his mind.

"Dad, you've traveled to some pretty wild places and had to deal with some rather extreme situations. You could teach people a lot by sharing what you've learned along the way."

Ted sat back in his chair and studied his wine as he swirled it around, looking thoughtful. Then a moment later, a big goofy grin wiped away his seriousness.

"I know what you're doing. You're just distracting all of us from whatever it is you're not saying."

Paul slapped his leg. "Touché!"

"What? No!" Anita reached for her wine and shrank a few inches in her seat.

"It appears the ball is back in your court, sweet Sis."

"OK, time to spill the beans!" Ted set down his glass.

"What beans?" Anita shifted uncomfortably. Then her coy smile broke into a monumental grin.

"Yeah, give us the goods, Anita." Mona winked at her daughter.

"This must really be good," Paul chimed in. "Come on, out with it!"

All eyes were on Anita.

"What is this, a family inquisition?!" Her arms were tightly folded, but her voice was ecstatic.

"That's exactly what it is." Paul's eyes twinkled.

Anita took a deep breath and closed her eyes. OK...no big deal...there's just this guy at work."

"Aha. I knew it!" Paul slapped the table. "You've been grinnin' like a fool all day I thought your face was gonna crack! Mine is aching just looking at you."

Anita waved a hand in front of her flushed face. She'd never met a guy worthy of such sweet torturous embarrassment at the hands of her own family.

Ted set down his wine. "That's great, Anita. Does he curl?"

"Ted!" Mona cried. But she couldn't help but crack up at her husband's innocent question.

"Yeah, can the guy sweep a broom like a maniac down a sheet of ice while chasing an oversized rock and screaming "harder, harder"?! Paul scrubbed his arms furiously, knocking the table and sending a knife and spoon clattering to the floor.

Anita just laughed, and then rolling her eyes glanced at her father.

"Of course, he curls, Dad."

"Good girl. Glad to hear it."

Quickly, before her family could drill her further, Anita set her sights back on her dad.

"So, Pops, about that book..."

"Book? What book?" Finished with his turkey and stuffing, Ted started in on the mandarin orange jelly salad on his plate.

Anita sighed, loudly, eyeing her brother for support.

"Your book!"

Now that he'd seen her squirm, Paul came to his sister's aid.

"It's true, Dad." Paul licked his greasy fingers, and Mona pretended not to notice. "Not many people can say they've dined and drunk with Kyrgyz government officials in their homes."

Ted looked up from his plate.

"Oh, you mean Duchaine, the half-crazed hockey fanatic." His hunched shoulders bobbed as he laughed. "The guy couldn't wrap his head around our love of curling. I remember one time he said, 'Why, you Canadians bowl on ice?'"

Ted shook his head, as everyone laughed. "But, man, was he totally obsessed with Wayne Gretzky."

"That night was such a riot." Mona winked at her husband.

They all wanted to hear the story again. Ted pushed his plate away, took a sip of wine then rehashed the old tale.

"Once in Kyrgyzstan, Mona and I were invited to dinner with Kumtor's environmental manager, Duchaine Kassanov, and his lovely wife, whom, as the story goes, he'd apparently divorced and married three times."

"Ha!" Anita folded her arms, leaning in, with a huge smile.

That hazy night long ago, after enjoying cocktails in the Kassanovs' living room, they had moved into the dining area, where a fresh, unopened bottle of Russian Standard vodka sat upon the table.

"In those parts, it was social custom to never leave the table before any bottle on it was empty."

With that, Paul comically reached for the wine bottle and topped up everyone's glass.

His family was all ears. Ted's face grew animated as he wove his amusing thread.

"One of the biggest honors to be extended by a Kyrgyz national was an invitation into their home, you see, so not wanting to offend them—and being a trooper, of course—I helped polish off every last drop."

"He was a trooper all right," Mona said, shaking her head slowly. The kids laughed.

Consequently, Ted had retained only two blurry memories of that liquor-fueled night of Kyrgyz excess.

"Number one," Paul raised a finger on cue. "What was it?"

"I discovered that Duchaine was an inordinately big hockey fan, so much so that he'd memorized the entirety of the 1994 NHL yearbook, and at one point, he even flipped through his old, worn pages reciting statistical errors."

Paul chuckled. "And number two."

Ted scratched his head, looking up at the ceiling. "Now, that one's a little hazier."

The girls giggled, sipping their wine.

"A drunk Duchaine perched precariously—foot up on his chair, bottle in hand—exclaiming to his sober spouse's chagrin that he loves his 'third wife' best!" Taking a sip, Duchaine had lost his balance, toppled over, and brought his wife crashing down with him.

"It was quite the scene," said Mona.

Everyone burst out laughing.

Then for a moment, Ted's face grew calm. His eyes drifted to the snow falling through the window, and he reflected on the significance of that cockeyed night.

"You see, in North America, with its impatient business society, work and pleasure mix like oil and water." Anita smiled as she watched her dad separate the wheat from the chaff.

"However, in that cold, turbulent new republic, different rules applied." He held his jaw, leaning his elbow on the table. "Two men bonded over the best vodka money could buy—not to condone excessive drinking 'cause it wasn't about the booze."

There was more to his story. When he'd returned to Kyrgyzstan the following spring, Ted made the kind of gesture that set him apart from so many other project managers and people in general. He brought his foreign friend a crisp, unopened copy of the newest NHL yearbook. Duchaine had been speechless, unable to obtain such an item in his own country. To thank his Canadian associate, Duchaine, not surprisingly, had reached into his desk and pulled out a fresh, unopened bottle of Russian Standard.

"Right then and there, we toasted to the Toronto Maple Leafs making it to the 1996 playoffs." Ted paused a moment, thinking fondly of Duchaine. "But here's the beauty of it: our personal relationship not only made the Kumtor project more enjoyable, despite its glaring difficulties"—he glanced at Mona, who nodded knowingly—"but it nicely greased business relations in a foreign country, and to top it off we're still friends."

"Nice," said Paul, taking a gulp of his wine.

Anita then jumped in, seizing the opportunity to make her point. "Now, don't you think that sounds like a very valuable lesson learned along the road of Ted Bassett's Great Adventures in Project Management?"

Ted laughed. "Oh, now it's great adventures."

Mona clapped her hands. "I'd say it was a lesson learned in when to swap your vodka out for water!" she teased.

Ted shrugged with mock guilt. "That's what your mother did."

"After dinner at the Kassanovs', your father was reeeally hurting that next day, and guess who was fetching him water then?"

Ted winked gratefully, at his wife. Her eyes twinkled in the warm cabin light. She'd been with him every step of the way.

"I won't argue that it's been a crazy adventure."

Being the wife and travel companion of a world-class mining executive was not for the weak of heart. She'd learned to love the ride, but Mona never knew which way the wind would blow them next. One year, home was the prairies of Saskatchewan. The next, it might be Toronto in the east or Vancouver in the west—or farther afield in Russia, Saudi Arabia, or the mountains of Kyrgyzstan. Sometimes the uncertainty unnerved her, but after thirty years of marriage, their partnership—not their coordinates—was the glue that held it all together.

"Sounds like one adventure worth writing about," Anita slid in, for good measure.

"I can't write. I'm an engineer!" Ted protested. "And not a very good one at that."

"No problem. Can't screw in a lightbulb? Find someone who can. You can hire one of them ghost writers," Paul suggested.

"See, Dad?" Anita nodded vigorously.

"Son, you sound like a project manager."

Paul chuckled.

Then Anita piped in with another reason that her dear father should pen his life's memoirs. "You know, you could probably fill a whole book writing just about Kumtor. Most people haven't had to deal with the horrors—and responsibilities—of an international incident like that!"

Ted's forehead rippled, and he stuck out his lower lip, blowing a stream of hot air up his nose. "You have a point there." Kumtor had it all: action, drama, suspense, comedy, and even tragedy.

Mona smiled gently.

Ted couldn't argue that being a project director on a multimillion-dollar globally scaled project wasn't your average gig—especially when something as big as Kumtor went down. Not one day went by that he didn't think about the life-changing events while working in Kyrgyzstan.

Paul's fist fell lightly on the table, interrupting Ted's thoughts. "I still can't believe you didn't call Anita and me to let us know you were OK!"

At the time, his daughter had been working at Kilborn Engineering's office in Saskatchewan. She heard the media reports that a helicopter had left a mine somewhere deep in the Kyrgyzstan's mountains and was still missing. Ted and his crew had reported the incident but at that time were prohibited from releasing any further details publicly. In the throes of the crisis, he'd even been unable to call Mona in Bishkek until several hours later. The next day, when Anita arrived at work after a fitful night's sleep, she was finally updated that her dad was not on the MIA chopper. She'd broken down in tears.

"What did you say again?" Ted turned to Anita.

Repeating the words, she muttered softly, "I still have a dad."

Looking at her father, Anita's hand floated to her chest. "I nearly had a heart attack."

Ted nodded, quietly. Suddenly he thought of his colleague and friend, Nick, who by some act of grace had been spared his life that day when he'd forfeited his spot on the flight due to a fear of flying in helicopters. Alas, someone else had taken Nick's spot. Ted's work had certainly brought him up close with tragedy. But it was behind him now. Why write about it?

"And then there was Bre-X," Mona piped in. "That nearly gave your mother a heart attack!"

Ted rolled his eyes, melodramatically. "It sure did!"

He chuckled at the memory of his eighty-two-year-old mother calling him up the day the world discovered the Bre-X gold mine in Busang, Indonesia, was a fraud. Aside from tragedy, Ted's career path had also brushed him up against the vile and corrupt underbelly of the business world.

"That's right. Bre-X," Paul nodded eagerly. "Hey, didn't you have drinks with that Guzman guy a couple of evenings before fingers pointed to him and he jumped out of the helicopter?"

"Or was pushed out," Mona interjected. "Who knows?!"

"Yes, I did," Ted replied. "Thing is, I actually thought he seemed like a pretty nice guy. You never know what's going on inside someone's head."

Michael de Guzman was the Bre-X geologist responsible for care and custody of the initial drill samples, the one who'd submitted the original estimate stating that the Busang gold deposit contained approximately two million ounces of gold. The world had bought in heavily. So when the truth came out, it rocked the whole financial system, and the mining industry was left in shambles.

"A record number of people left the field—and no doubt we'll be recovering for years to come."

Feeling his daughter's eyes on him, Ted beat Anita to the next punch. "But, you know, tons of books have already been written on that topic. Bre-X is done and gone."

Anita's eyes shone, undeterred. "True, but no one's quite had your unique experience. I mean, didn't you take the fall for someone on your team who should have caught the fake gold samples?"

"Not really. Once the fraud was discovered we went back and reviewed all the data, and hindsight told us that the assay data was too consistent from drill hole to drill hole. This is not something the team would have been looking for while developing the geological model."

Mona's fork stopped midway to her mouth. Years had passed since the incident, but the mere mention of Bre-X could still make her stomach tighten. Instinctively, she empathized with her husband. "That may be so, but in the corporate world when something goes wrong, someone has to take the blame."

Ted's face softened as he looked at his wife appreciatively, but his voice remained matter of fact. "True. Bre-X changed the direction of my career, that's for sure. In good ways, and in bad."

Tapping her fingertips together, Anita grimaced. "Hmm...sounds intriguing...I'd read about it," she winked at her dad.

Paul smiled at his sister, thoroughly amused by her persistence.

There was no question Ted had been through the wringer in the last few years. First, there was Kumtor. Then he'd been slammed by Bre-X, and then cold, unforgiving Siberia. Actually, it was a surprise he had any hair left on his head. Suddenly, Ted's forehead relaxed. He looked around the table at his family. Life was good and full of blessings. He could sleep at night, despite everything, because he had a clear conscience.

"Your mother—what did she say, again, when she found out Bre-X was a bust?" Mona prompted Ted, with a smile.

"Oh, God, she said, 'Ted, you are in deep shit now!'"

Everyone broke out laughing as they pictured Granny cussing. Even she had bought into Bre-X gold and was righteously peeved when it turned out to be a scam.

"What did you tell Gran?" Paul asked, folding his hands behind his head.

"I told her the truth. 'Don't worry, Mom. I personally have not done anything wrong. I was just the guy in charge.' It all worked out in the end."

"But you didn't get the promotion." Anita looked at her father with empathy.

"No, I did not, Anita. Pierre got VP. I got Siberia."

Anita shrugged her shoulders up to her ears. "Sounds dreadfully cold."

As they all knew, Anita only put up with the cold if it was in a curling rink, anything for the love of the game. Like many a young Saskatchewanite, she'd thrown rocks since she was nine years old, but Siberia was not a place she could ever see herself going.

"Actually, Russia wasn't half bad. The project was a dud, but the people and country were great," Ted replied. "And the vodka was darn cheap!"

Wielding a toothpick, Paul snickered.

"And finally, after all that, I got my great success." Ted had their full attention, so roguishly, he changed the subject. "But enough about me. Tell me, Anita, how's living in Prince George? And who's this fine fellow you've told us so much about? And more importantly, when do we get to meet him?"

Tables turning, Anita shrank back in her seat blushing again.

"You didn't think we'd forget that easily, did you, Sis?" Paul chirped in.

Playing along, Mona turned to her daughter. "Now, honey, just tell us his name, where you met, and whether he is tall, dark, and handsome, curls skip or second, and then we'll leave you alone. We promise."

Anita laughed at her mom and came unraveled. "OK, OK." She pushed a piece of hair behind her ear, her face bright pink, and her voice rose a pitch, quivering slightly as she spoke. "His name's Mike. Mike Cochrane. And yes, he's all three!"

Paul sat up in his seat, gob smacked. "My God, she's actually totally smitten!" Anita had dated before but never had a serious boyfriend, and he'd certainly never seen his sister this wound up over anything or anyone.

Anita cleared her throat. Now she was on a roll. "We met at KPMG. He's an account manager there. He asked me to lunch, and it kind of went from there. Then he invited me to join the Prince George curling league, and we started running into each other at the club. Man, he plays a good skip. Dad, you'd be impressed."

Ted narrowed his eyes. "Good enough for Team Bassett?"

Anita nodded decisively.

"Well then, you should invite him home for Easter!"

"Yes, great idea!" Mona cheered.

"Yeah, bring him around." Paul raised a glass.

Anita smiled quietly, eyes drifting to look over her right shoulder toward the kitchen. "We'll see." Then she popped up from her seat and timidly made a move to clear the table. "Who's for dessert?"

Paul blew the whistle. "Trying to slip away, are we?"

Anita sighed and looked to her dad for help, having had enough of the spotlight. "Pretty sure we're all done with me. Now, about that..."

Ted came to his daughter's rescue. "Book! Yes, it's not a bad idea, after all. How about—when you're ready—you consider bringing this fella Mike home. And I'll consider penning my great, adventurous memoirs—one day."

Anita beamed at her father.

Ted scratched his head, making a mental list. "Now let's see—to write about the principles of project management—well, you've got to have scope and definition—a budget, a master schedule."

Mona rolled her eyes. "Ugh, that sounds a little boring."

"Not quite what we had in mind." Paul feigned a yawn.

Leaping to her dad's defense, Anita sat back down. "But, you see, it doesn't have to be boring!" Anita launched into an impassioned plea. "You've been touched by all of these people and experiences around the world—different cultures and politics. You've seen the nitty-gritty of your industry and become a leader in your field! Besides, I'm not even a project manager and your stories, over the years, have taught me so much, Dad—about work ethic, leadership, being a team player on and off the ice."

"Actually, me too, come to think of it." Mona leaned in, smiling at her husband.

Paul shrugged. "Me three."

"Really?" For a moment, Ted sat back, looking at his family, stunned. "Well, when you make it sound like that..." He pretended to brush lint from his shoulder. He'd had no idea. Perhaps a book wasn't such a crazy notion, after all.

"Who knows, maybe it'd even be a best-seller!" Mona leaned back in her chair, stretching her legs.

Paul crossed his arms, and then furrowed his brow. "But do engineers even read?"

Anita clucked her tongue. "I don't know, do they?"

"Oh, yeah. We're avid readers." Ted laughed. "There's nothing like a good operations manual or balance sheet or safety management checklist to sink your teeth into before bed."

Anita held her hands in her head. "OK, maybe you will need a ghost writer!"

Where had the time gone, Ted wondered, suddenly, looking around the room. For a moment, he saw in Anita the wide-eyed child who'd greeted him eagerly at the front door after his long business trips. Without fail, he'd hand her a doll from every different country he visited. When she was a bit older, he'd still bring home the doll, but it was his stories and adventures she longed to hear. He didn't realize he'd left such an impression.

Ted reflected a moment. "I don't know...I can't imagine having time to do such a project. Maybe when I'm retired."

"Honey, you'll be writing that book when you're ninety!" Mona cried.

"Not ninety. But maybe seventy-five," Ted calculated.

Anita beamed, looking victorious. "So you'll consider then..."

"Only if you'll bring this Mike fellow home to meet your crazy family. He sounds like a neat guy."

"Ahhh..." His daughter backpedaled.

Mona shook her head. "Ted, you have no mercy."

Father and daughter faced off. Arms folded, Anita tapped her foot melodramatically.

"You know, you're very persistent, maybe you should have been a lawyer," Ted teased.

"I'll think about inviting Mike home for Easter. Now, that book?"

Ted rubbed his chin looked up at the ceiling for a moment, seriously considering the possibility. "For you, one day, I promise, I will!"

CHAPTER TWO

Commitment

Ted's story starts with a ring.

The Iron Ring is a simple, circular band worn by professional engineers. New graduates receive this ring—a symbol of strength, quality, and integrity—after swearing to uphold these tenets in work life during a ritual ceremony whose origins have faded from collective memory. First conducted at Canadian universities in 1922, the United States has now adopted the Iron Ring tradition at many schools, following in its northern neighbor's footsteps. The distinguishing ring is worn on the pinky finger of the dominant hand to serve as a steady reminder of the engineer's commitment. Just as a marriage begins with a promise, rites and rubric matter little compared to the seemingly trivial day-to-day actions, from that day forward, which can either exalt or undermine the bond. For better or worse, new engineers vow to use their skills and knowledge to engage only in honest enterprise and to wield nature's finite resources in the name of growth and progress. Thus the work begins.

Amid "official" facts surrounding the Iron Ring, its origins and rites are stories of cloaked rituals, collapsed bridges, inebriated after-parties, and non-programmable calculator Olympics. Legend has it, for example, that after a night of excessive drink (which the intrepid engineer can accomplish with great panache) the Iron Ring will turn a wee finger black. In other lore, the ring is meant to serve as a sober reminder of engineering's past monumental failures.

The collapse of the Quebec Bridge during its construction in 1907 claimed the lives of seventy-five workers. A public inquiry into the mishap found that faulty design and poor planning were to blame. Then a further foul-up during its rebuilding in 1916—a glitch in a hoisting device—sent the bridge's midsection plunging into the St. Lawrence River, killing thirteen more people. The world's biggest bridge-building disaster spurred a public debate around the role that engineers play as civil custodians in modern society and the need for stiffer regulations. Within five years, most Canadian provinces had passed licensing laws for practicing engineers.

Seven prominent engineers, who called themselves the Corporation of the Seven Wardens, Inc., met in 1922 and decided that engineering needed its own version of the medical profession's Hippocratic oath; the purpose of such would be to unite engineers under a common code of ethical conduct and impress upon new initiates the weight of their social responsibility. The Seven Wardens commissioned Nobel laureate writer and poet Rudyard Kipling, who designed the Iron Ring and wrote "The Ritual of the Calling of an Engineer," a clean, reverent pledge to serve the public good and uphold only the highest professional standards.

> The Ritual of the Calling of an Engineer has been instituted with the simple end of directing the young engineer towards a consciousness of his profession and its significance, and indicating to the older engineer his responsibilities in receiving welcoming and supporting the young engineers in their beginnings.

The rest is history, but like any story—depending on who's telling it—the message may get a rewrite. Curiously, no institution conducting the Iron Ring ceremony today bothers to draw the link between a rich tradition and its origins. In fact, many schools outright deny the "myth" that the Iron Ring is meant as a reminder of the deadly Quebec Bridge disasters. Whereas science learns from its mistakes, modern public relations has all but blotted out engineering's inglorious past. But history holds the facts, for anyone wishing to interpret them.

For starters, two of the original Seven Wardens of the Iron Ring were chief engineers on the Quebec Bridge project. What else could have motivated these men to try to safeguard their profession? Further fodder for the ring-reminder theory: during an official Iron Ring ceremony, the officiant will tap out a message in Morse code using a hammer and anvil. This hammer dangles from a chain, which circles the room, roping in the new initiates. Suspended from one of the chain links is a dangling memento: an iron rivet from the Quebec Bridge.

As for the eighty-eight men lost in the Quebec Bridge accidents, thirty-three were Mohawk ironworkers from the Kahnawake reserve near Montreal. They were buried at home under crosses poignantly fashioned out of steely gray-blue girders.

The original collapsed bridge beams still rest in a watery grave at the bottom of the St. Lawrence River—a heavy load, forgotten by some. That's as much a part of engineering's legacy as the new bridge spanning the river today. The Iron Ring is a reminder of human fallibility and the will to recover and raise the bar higher.

* * *

As Ted flew in a 737, bound for Almaty, Kazakhstan, he twisted the beveled band on his right pinky finger. The confident senior executive of the Mining and Metals Division still considered the day he got his Iron Ring to be the true beginning of his professional journey. It was February of 1993. Ted was forty-six years old. He had nearly two decades of solid project work tucked under his belt. But when you're soaring high above the clouds, in an endless expanse of blue with no ground visible, it's easy to lose your bearings and get knocked off course.

Ted was a top dog on the largest gold-mine project in Southeast Asia. By then, he thought he'd seen it all. But the Kumtor project would test him in many ways, professionally and personally, exposing his blind spots and forever shifting his internal compass. Ted's journey had just begun.

CHAPTER THREE

Sleeping on the Ore

The beat-up yellow-and-white Volkswagen van bounced across the rugged Kyrgyz landscape, from Bishkek to Balykchy. The bumpy ride reminded Ted of peddling a long-forgotten go-cart down the dusty roads of his Saskatchewan youth. As they jounced and wobbled into the third hour of their journey, traditional Kyrgyz folk music playing on the state-run radio station drowned out only by intermittent chatter, Ted wondered if the van's shock absorbers had been tuned anytime in the last decade.

If their old-world hippie bus broke down leaving them stranded in the middle of early post-Communist nowhere, he was sure they'd figure out something. But any delay would put their schedule on the line—which would make for an unhappy client. Ted drummed his fingers on the steering wheel to a tune reminiscent of a Ukrainian polka. Sitting beside him, head stuck in a map, was his chief mining engineer, Gord Clark. And behind him sat their client, Mike Lindberg, who always appeared to be tied in a severe knot. Art Carino, their metallurgical expert, was sprawled out on the backseat.

Gord's lips moved, silently, as his index finger scanned the roadways. Their route was fairly direct, but Ted felt confident with Gord as his navigator. Back in 1988, he'd snatched up the talented and soft-spoken Vancouver native and prairie transplant for his team within hours of a key industry merger that'd left

Gord out in the cold. As Ted cruised along, he recalled the day, four years ago, when El Dorado Nuclear had joined forces with Saskatchewan Mining Development Corporation (SMCD) and gone private, forming one of the world's largest uranium outfits, Cameco Corporation. Whenever something went down on Saskatchewan's mining scene, everyone knew about it before sun up. That day, Gord had gotten the call at eight in the morning to be out by four.

Ted glanced to his right, at his even-tempered colleague who'd taken the news in stride. Shake-ups to management rosters were all part and parcel of corporate mergers and acquisitions. It was nothing personal. Cameco's loss had been Ted's opportunity. And now Gord worked with Ted and Cameco was their client. Feeling Ted's eyes upon him, Gord looked up from the map.

"There's only so many roads out here, and one of them will take us where we need to go. I think we're on it."

Ted chuckled. "As long as this beetle keeps up, we should be there by one."

The geological engineer set aside his map and then flipped open the glove compartment, shuffling through its sundry contents. He pulled the Tragically Hip's album *Road Apples* out of a sleeve.

"Mind if I...?"

"Not at all."

Gord gratefully slid the compact disk into the retrofitted stereo system he rigged up for their ride. Little Bones started blaring.

Before he'd even packed a box, two out-of-province consulting firms had already solicited Gord. Not crazy about the idea of uprooting his young family from Saskatoon, he'd decided to hold off on making any rush decisions. But when Ted rang him up the next day, Gord had had a change of heart. He didn't even hesitate. Having worked on projects with Saskatchewan-based Kilborn Engineering in the past, Gord had heard only good things about Ted and the company's president John Somers. They were the kinds of guys you wanted as partners. In the mining industry, you had to be careful who you rubbed shoulders with lest someone else's dirt rubbed off on you. Ted and John were squeaky clean.

Gord adjusted the volume, and his head started moving to the beat. Ted joined in, bobbing his chin in and out, eyes on the road. He felt energized for the miles ahead. From the backseat, Art belted out lyrics "two fifty for a hi-ball and buck and a half for a beer." Even Mike's face twitched a little, and he uncrossed his arms. How could you resist the Hip?

So here they were; driving across a bumpy road through the mountains of Kyrgyzstan, an elite team of corporate outbackers on a time-sensitive fact-finding mission. At the end of the rail line, a five-ton pile of quarried rock was waiting for them. There, they'd find sledgehammers, hard hats, and a Kyrgyz contractor by the name of Altin Jyrgal uulu, who just that morning had surprised them by claiming to have transported the ore samples from the Kumtor mine site. In reality, the rock could be from anywhere, and it could be salted with "gold dust." Tampering of ore samples was always a distinct possibility. You could never be too careful when it came to precious gold, so little of it could gild pockets, corporate coffers, and entire economies and taint all manner of people.

Ted and his team were on guard. Just two days back, they'd spoken with Mr. Jyrgal from Bishkek via a spotty cellular connection. He'd initially told them that the dirt road into the remote mine site located in northeastern Kyrgyzstan was impossible due to a late-winter squall, which had dumped several feet of snow inside the Tien Shan Mountains. As it had appeared, they were shit out of luck.

That's when Mike Lindberg's face had started turning red. Cameco's VP of projects on Kumtor—and Ted's client—wasn't a delight to work with on a smooth day. He was hard on people, impatient by nature, and constantly worried that someone would take advantage of him. Through the rearview mirror, Ted glanced at Mike who was shaking out his arms, and then he inconspicuously loosened his collar. Ted turned to Gord and signaled him to crank the volume another notch. Gord winked as he complied.

Ted reviewed the situation in his head. If they couldn't get in there and get the ore on this trip, they'd have to wait until they returned to Kyrgyzstan in June, which meant a three-month delay on metallurgical testing. Such a holdup would throw Kilborn's and Cameco's schedules out of sync and delay contract negotiations with the new republic of Kyrgyzstan on the Kumtor Gold Mine. Fortuitously, bright and early the next morning, out of the blue, the Kyrgyz contractor had called them back to say he had their rock. When the call came, Ted and his team had been wading in contingencies, discussing how to proceed on their feasibility study without any actual samples, subduing Mike in the process.

"Fan-friggin'-tastic!"

Mike had rapped the table, diffusing himself like a cherry bomb, while the others had exchanged glances, silently considering their options in light of this new development.

Apparently, Mr. Jyrgal, noting their urgency, had made the trek with a crew through deep freeze and snow into the mine site, "which he knew like the back of his hand," having worked up at Kumtor with the Russians preindependence. Shovel by shovel, through the night, they'd loaded up his dump truck and then transported the heavy cargo more than two hundred kilometers west to Balykchy, according to the contractor.

Ted scratched his head, considering their situation, as he drove on. Could they trust this Altin fellow? They were riding a thin, bumpy road of calculated risk. The samples would be used to make critical calculations and decisions around the ore grade, metallurgy, and processing techniques. They must not be compromised. Any whiff of scandal or imprudence resulting in nonconfidence of partners was corporate suicide. Millions in public shareholder investments were at stake. Their reputation, like the gold ore, was both valuable and finite. They had to be careful with whom they dealt. And so they'd set up a meeting for that afternoon, grabbed their work boots and some tools, and hopped in the van for the five-hour journey. Their joint decision was to proceed with caution.

The song ended. Breaking the silence, from the seat behind them, Mike piped in with his lingering concerns, beating a dead horse. They'd already hashed through the situation that morning. "I sure hope this Altin guy is as good as his word." His mouth was downturned slightly, and he was starting to sweat through his cotton shirt.

"No kidding," Art piped in, from the back of the van. The good-natured Filipino chap had spent the last half hour gazing out the window at the vast panorama, apparently lost in the moving folds where mountain met valley and the occasional Arabian horse grazed peacefully.

Ted met Mike's nervous eyes in the rearview mirror. Just like their rock, Altin was an unknown quantity. None of them in the vehicle had ever met their zealous delivery boy, although as an employee of Cameco's subsidiary outfit Kumtor Operating Company, Altin technically worked for Mike.

Holding Mike's gaze Ted spoke affirmatively. "Let's meet this guy and see where we stand."

Mike half nodded in agreement and then looked away, clearing his throat, and said nothing.

A moment passed, during which Ted watched his client's chest expand, and then in his usual manner, he brought a pessimistic outlook to the circumstances. "Our Altin, he could be, I don't know, a KGB agent, working in cahoots with the Kyrgyz to get us to sign on a lemon. He worked for the Russians before separation, didn't he? He could be leading us into some trap. This could be an elaborate set up for a stock scam—delivering us tainted samples in the middle of the badlands. God knows what from God knows where."

He heaved and sighed. The others shifted in their seats, unresponsive. Better to let him burn off steam than engage in his agonizing line of conjecture.

Scams happened. Half-truths, fabled moose pastures, get-rich-quick deals. Mike wasn't all that off base with his paranoid suppositions. But that's why they were going in with their eyes wide open and ears cocked for bullshit. It was true, the Russians had once had vested interests in Kyrgyz gold. During the Communist empire, the Kumtor mine had been explored extensively by the Soviets. But when Kyrgyzstan gained its independence in 1991, the sovereign nation took control over its natural resources.

The Saskatchewan mining team now had in their possession all of the Russians' original maps and geological information, albeit in a foreign tongue. But they couldn't just accept the old data; they had to verify it themselves. Science had to be on their side. Dubious or unsubstantiated core samples wouldn't fly with the public investment community. And haste, lack of hard data, or blind faith would unnecessarily open them to a poor investment, dubious partnerships, lawsuits, scandal, or simple guilt by association. They could leave no room for doubt. But brooding was pointless.

Mike gripped the side-door handle as he spun his web of doubt. Ted just nodded, waiting for his client to be done.

Risk and uncertainty were a part of every project, even with standards and protocols in place for everyone's protection. It wasn't that Ted was free of apprehension or doubt, but you could learn to live with the questions and anticipate and minimize potential problems and threats—or let it all give you a coronary.

Mike paused and Ted took the opportunity to cut in. Sometimes it was simply a matter of changing the subject. In almost all situations, Ted found that having a sense of humor did wonders.

"Hey, Mike—did you hear the one about the pastor, the doctor, and the engineer?"

"What? No," a tight-lipped Mike replied, still staring out the window.

Ted launched into an old favorite, reciting it like a seasoned pro. "A pastor, a doctor, and an engineer are on a golf course behind an especially slow group. When the marshal comes around, they decide to ask him what the deal is."

Gord looked up from his map, all ears, and Art leaned over the seat back.

Keeping his eyes on the road, Ted continued. "He tells them the slow play is because it is a group of blind firefighters who saved the clubhouse from a fire that blinded them, so they get to play for free."

Gord cracked a smile, and Mike looking mildly annoyed by the convivial hold on his attention. Ted continued. "The pastor proclaims, 'That is terrible. I will say a prayer for them.' The doctor says, 'I can contact an ophthalmologist friend who has done wonders with the blind.'" Ted raised his pitch for the punch line.

"Then the engineer asks, 'Why don't they just play at night?'"

Bada bing, bada boom. The van rocked with laughter. Mike broke into a reluctant smile and, for a moment, forgot his beef.

"You got any for us, Mike?" Gord swung around in his seat toward their client. The VP of projects rubbed his hands on his pants, unable to resist a good-natured challenge from a fellow engineer. So he lobbed one back.

"OK, so a manager is on a hot air balloon and an engineer is on the ground."

"Uh-huh." Art leaned closer, and Ted eyed Mike in the rearview mirror in a supportive manner.

"The manager is lost and yells to the person on the ground 'Hey you, do you know where I am?' The engineer replies, 'You are on a hot air balloon approximately ten meters from the ground heading north-northwest at about one mile per hour. The manager, irked, shouts back, 'You must be an engineer, your answer is correct but it doesn't help a bit!'"

Gord smiled expectantly. Mike looked like he was enjoying himself. "The engineer snapped back 'and you must be a manager. You got where you are by hot air, and it was entirely your fault that you're lost, but somehow it now sounds like it's my fault!'"

Mike snorted, and Art did a drumroll on the polyester seat back.

"Ha! Nice. Good one." Gord chuckled and then shook out his map like a broadsheet newspaper and got lost again in the reliefs and roadways. Levity once again returned to their journey.

Before Mike could withdraw again, Ted kept the conversation flowing in his direction. "You getting hungry, there?"

"Sure am!"

Ted pointed toward a town just a few clicks off. "Maybe we can pull in for a bite."

"Sounds like a plan. Hopefully they've got something to eat in Kyrgyzstan besides eggplant."

"Or mutton roadkill," Art called from the back of the van.

While in Bishkek, they'd hired a Russian coordinator named Igor after discovering that none of the restaurants open in town actually had any food to serve. So for weeks they'd been at the mercy of Igor's culinary tastes. Oftentimes, if it wasn't yet another creative rendition of the abundant nightshade vegetable, they weren't quite sure what they were eating. Maybe best not to know.

Whatever lay ahead, Ted was sure they could handle it. Whether this Altin guy could be trusted was a matter of utmost importance. And so, their job—apart from obtaining the core samples and guarding the ore until it could be shipped home for testing—was to connect with Altin and read the man closely. This kind of literacy wasn't taught in engineering school. You had to learn it on the job. Ted adjusted his glasses and resumed focus on the road ahead.

A few minutes later, as they pulled over and got out of the van to stretch their legs, Ted let out a quiet sigh. His stomach groaned as they headed toward the diner. Worrying about his client, Ted hadn't realized he was so hungry. It was time to fuel up for the work ahead.

* * *

They'd been slogging away for three hours. Six men drenched in sweat and wielding shovels tackled a four-by-ten-foot pile of rock, one boulder at a time. They took turns, making headway with a couple of sledgehammers borrowed from the Kyrgyz and Russians, along with some rusty, old shovels they'd found on-site near the rail yard. Even Altin and his colleague, a Russian named Vlad, were in there, helping get the job done. They'd arrived in Balykchy that afternoon and could have easily just dumped the rock and headed home. Instead, in light of the apparent urgency to get the assay sample off to lab, they insisted on sticking around to help break up the ore.

Ted's team was grateful for the assistance. Although they were working in a relatively flat zone, located at the outskirts of town by the west end of Lake Issyk Kul, they were still 5,318 feet above sea level. This was rugged work made worse by the higher altitude; their bodies, unused to the altitude, were using oxygen less efficiently. Their helpers had come prepared with warm layers, high-electrolyte drinks, and supplemental oxygen tanks.

It was a beautiful afternoon with a temperature of minus 5 degrees Celsius (23 degrees F). The sun was shining in a cloudless blue sky as it arched slowly back toward the mountains. They had three hours left of daylight, and their work cut out for them.

Art wiped his forehead and whistled and then crouched down for a handful of snow. He was sweating through his thermal shirt but was working too intensely to catch a chill. He took a deep gulp of air.

"Man, it feels like the Bahamas in February."

"No shit." Mike grinned, red-faced and relaxed. A little hard labor could transform a man right before your eyes.

Gord was on a mission, getting ready to start up the hammer again to break down a huge boulder. He didn't stop for small talk. Ted was also loving it, sleeves rolled up and jeans covered in dust. He always felt a little badass in his steel-toed work boots. He shook the rocks from his shovel and then dug back into the pile to reload. Days like this put the "man" in project management, Ted thought.

Although he would happily work beside anyone cut out for the job, there still wasn't much diversity in the mining field. In 1993, the pipeline of skilled female talent tended to dwindle before advancing to the executive ranks. You had to have thick skin, a yen for travel and adventure, shrewd business skills, and a willingness to get your hands dirty—all gender-neutral qualities. Indeed, over the last decade, there'd been an increase in female geology and engineering graduates, but men in their fifties and sixties were still making the hiring decisions. And so, typically, trips like these were fraternal endeavors.

In between loads, Ted side-glanced over to Altin and Vlad, who'd barely paused since they'd dug in. They were moving rock like it was play-box sand. Altin was a mild-mannered man with dark, curly hair, a medium build, and calloused hands. He'd looked Ted straight in the eyes when they'd shook hands earlier and had bowed his head ever so slightly in a manner both humble and sincere. His Russian comrade was an affable and intelligent fellow, with pipes the width of the Volkswagen's tires. He was a stone-pushing machine. If they kept up at this rate, they'd be done before sundown.

So far, Ted had a good feeling about these guys.

Over the next few hours, they cut, coned, and quartered the rock. This was a standardized method of obtaining a representative ore sample. Altin and Vlad respectfully stood off to the side during this part of the process. After breaking the boulders down into smaller, manageable pieces, they piled the rock into large cones, mixed it up, and then divided it into quarters. Only one of these quarters would be sent off for metallurgical testing in Canada. Their results would be compared to the Russian samples and would give Cameco and its partners a clearer, more reliable picture of the ore and what to expect when they started processing it. Other testing would be required. But if the ore made the grade, the project was a go.

***

"But isn't a gram of gold a gram of gold?" Vlad asked later, a bit perplexed, as the guys sat around like roadies inside the Volkswagen van, passing a bottle of Kyrgyz cognac that Vlad had pulled out from a satchel when their work was done. They also snacked on some smoked chicken, which, when the remnants were inspected later in the light of day, hadn't been thoroughly cooked. Everyone was exhausted but content to kick back inside their cramped beatnik quarters, warm up, and share a drink.

In an almost fatherly tone, Gord spoke. "The bottom line, fellas, is can you pull it out of the ground and make a buck?"

The Soviets had yet to convert to the standards embraced by the international investment community. Gord elaborated. "These days, the true value of gold is tied to many extrinsic factors: the ore's quality and concentration, how easy it is to access, the capital costs of the project, current market values—you name it."

Gord went on, waving his hand, sounding a bit like a textbook, but the knowledge he peddled came from first-hand experience. "Historic trends, basic supply and demand—all taking into account any global sociopolitical variables, which could assist or impede any stage of the process. The considerations are numerous and complex, but in a nutshell, anything less than a fifteen to twenty percent return on your investment, and the gold stays in the ground."

Vlad and Altin nodded their heads vigorously. They were like two hungry fish who'd suddenly found themselves in a pond larger than they'd ever swum in before, and the pond was full of delicious minnows!

Altin took a swig of cognac and shared his philosophic view. "Kyrgyzstan, my country, she is lost. For years, my country was controlled by the USSR. When they broke up, and we became an independent state, we were no longer Communist, but we do not know yet what it truly means to be capitalist."

Vlad nodded and eagerly jumped in. "In Russia, we trade gold for oil or apples. Gold is gold. We know nothing of these market forces and 'cut-off grades.' For the Russians, if there's gold in the ground, we dig it up. Why not? We have no real competitors."

Vlad went on to describe how, for decades, in his country, gold had been classified as a strategic mineral. "Russia did not want anyone to know what we were mining—or how much of it. This was all...as you say...top secret."

The corollary to this secrecy, however, was extreme isolation for Russia's top engineers and PhDs, who weren't able to attend the mining conferences in Toronto and New York, share their research and findings, and kibbutz with their international colleagues. If someone invented a better mousetrap, it was lost on them; conversely, what the Soviets knew, for example, about working in high altitudes or frozen subterrane extraction methods went unshared with the rest of the world—a disservice to all.

"But now that we are an independent state, we can finally connect and share with the rest of the world. That is a reason to celebrate." With a warm smile, Vlad made a high salute with the bottle.

"Here, here!" A slightly tipsy Mike raised a hand in tribute.

The bottle was passed off to Ted, who took a sip of the stiff drink and grimaced. Every man had dirt and dust encrusted in his pores. Hard work was like bonder's glue; it gelled people together.

They were disheveled, tired, and yet uplifted by the company and spirits, content to chin wag and squat in a bus under the stars in an ancient land that was just awakening to a new and uncertain era.

Then Art bowed ceremonially. "Great to work with you guys." He spontaneously recited a portion of the "Calling of the Engineer": "To dignity, integrity and knowledge." The others joined in the toast, even Altin and Vlad, who weren't familiar with the western oath of professional engineers but liked the sound of it.

Vlad then spoke up eloquently. "In old Russia, we pledged our duty to the Soviet government, to practicing the principles of communist morality. Now, thank God, we can pledge our allegiance to each other...and to science." Altin nodded to his friend.

Gord took a drink and raised the bottle one last time. "To science!"

"To science!" They all chimed in.

Despite their seeming differences, these six men of varying backgrounds had much in common interest, and they chatted into the wee hours of the night. Finally, Altin and Vlad crept back to their truck to catch a few winks before the drive home to Bishkek. Ted's crew stretched out in the seats of the van as best they could for a bit of shut-eye.

Mike nodded off to sleep within seconds, a rolled sweater lodged between his head and shoulder. Gord and Art were soon to follow, snoring in symphony.

Ted checked his watch, yawned, and stashed the burlap bag containing their chosen rock sample tightly between his feet. Now that they had it, he couldn't let the rock sample out of his official "care and custody," or like airport baggage left unattended, it could be deemed compromised or unreliable and tossed, making their trip and efforts wasted. So he slept on it.

Thankfully, after all, Altin's word would turn out to be as good as gold. But he was only one man. Ted really had no clue what the Kyrgyz or Russians were capable of, given the opportunity. So he left nothing to chance.

Three hours later, when the golden sun rose over the Tien Shen Mountains, Ted rubbed his bleary eyes and stretched his sore muscles. After some instant coffee, the men bade their two new friends and foreign work companions goodbye and pulled back on the road. It was time to ship some rock home to Canada.

＊＊

Altin's rock made the grade. But all important data had to be confirmed; they could make no assumptions. So, that spring, on the team's next trip to Kyrgyzstan, they did their final round of sampling to confirm the accuracy of the Russians' initial metallurgical data. This involved duplicating the Russians' unique methods of collecting samples, which meant gaining access to many kilometers of remote underground tunnels, dug in the 1980s during the mine's initial exploration phase.

First, a Kyrgyz-based Mine Safety Team, along with Kumtor and Kilborn personnel, made the challenging uphill trek across a frozen snow field at an elevation of more than four thousand meters to inspect the geotechnical integrity of the old tunnels, its water levels, air quality, and flow. After socializing the night before, the Kyrgyz made it to the mine site first and were sitting at the entrance smoking their cigarettes when their Canadian counterparts finally arrived, out of breath. The Kyrgyz offered them oxygen, but no one needed it.

Due to a tight schedule, a back-up order of Canadian fans and ducting had been shipped to Kumtor. But the safety team reported the tunnel's natural ventilation was enough for personnel to work safely, so the equipment never made it to the site. Wielding diamond drills, a Kyrgyz work team worked for three days at high altitudes in the underground channels, collecting samples from three different levels. The sampling program was going so well, in fact, that supervisors thought they'd finish the next day, ahead of schedule. But the following morning, the crew leader told management that they were not going back underground.

At first, Gord and Ted were confused. They thought their team was quitting on them until the crew leader explained that three days at high altitudes were enough.

"Today we build boxes at base camp. Tomorrow, we finish drilling," he stated plainly.

"Boxes? What do you mean?" Gord scratched his head.

"To ship the rock samples in, back to Canada."

So on the fourth and fifth days, the Kyrgyz built wooden boxes, and on the sixth day, they returned underground and finished the job. The samples were packaged in the homemade boxes and shipped for metallurgical testing. The final test results confirmed the Russians' initial ones. Based on this data, they were able to calculate the ore reserves at Kumtor, and the project got the green light.

LESSONS LEARNED
1. In mining, care and custody of the samples used to determine the grades, metallurgy and processing techniques must not be compromised.
2. All important data used must be confirmed. No assumptions.
3. In almost all situations, a sense of humor helps.

CHAPTER FOUR

On Being Principled

Jim sat at his home-office desk, feet firmly planted on the cement floor. He hadn't lifted his head in over an hour, immersed as he was in a schematic drawing of a pair of twin trellised gates, which he planned to build later that summer with his next-door neighbor, Matt. The designs were almost complete.

The semiretired seventy-eight-year-old consultant eyed the precision 2B-pencil blueprints with the intense focus of an eagle. Then, as he perched over the domain of his neatly piled surroundings, having already devoted part of the morning to the Panuke Field file and several chapters of former US Secretary of State Colin Powell's latest book, Jim finally looked up, adjusted his bifocals, and surveyed his surroundings.

For the third time, he'd failed to acknowledge his wife calling him upstairs. Above his head, Beryl was characteristically bustling around getting ready for their guests. He could hear her moving from pantry to linen closet, up the stairs, down the stairs and then back up again, making a dull racket like a baby elephant in a china shop. He found the cacophonous sounds of her endless activity comforting, while his cerebral introversion could sometimes still unhinge her. For sixty years now, she'd crinkled up her nose in that cute way when he teased her.

Jim needed Beryl like he needed his extended solitude and pocket protractor—so elementally, that neither a civil war in Turkey nor a raging storm over a heaving rig in the Atlantic could keep them apart. The rich fabric of their friendship, marriage, and lifetime together bound them, like an intricate web of connective tissue woven around a human skeleton. He could smell her amazing strawberry-rhubarb crumble baking in the oven upstairs, and the pang of hunger split the atom of his fierce focus.

Jim was a deeply loyal creature of habit and a master of toil. Every morning was the same start. He awoke at 6:00 a.m. to a cup of coffee and some toast with jam and read the business section of the newspaper before retreating to the basement. There, he'd remain for hours, lost in the dusty bowels of their two-thousand-square-foot subfloor space, which Jim had converted with typical zeal into a vast storage mecca that, still to this day, his wife showed off proudly when giving new houseguests "the royal tour." The extremely low ceilings made the space fit for a Hobbit, Beryl joked, proclaiming that the basement was the only place on earth where she didn't mind being short. Then she'd add, to Jim's mock chagrin, that his shoulders were starting to slump from his extended periods of underground "navel-gazing."

As Jim straightened his back, he felt his muscles revolting and rubbed his neck.

"Down here, Squirrelly!" He hollered, clearing his throat. Where else would he be? Then he shuffled some books and papers around and glanced at a snapshot of Beryl and himself in a brass frame, taken in Hawaii over three decades earlier. Beside it was a faded photo of Jim looking sheepish in a pink tutu and undershirt at one of his company parties back in the day.

He tapped the desk with his fingers, in a restless, thoughtful way. Finally, his eyes rested on a family photo, a group shot of all of his children and grandkids, piled on the back porch, taken that last summer. Jim's eyes lit up. Upstairs, the doorbell rang.

He checked his old Seiko watch, a gift from an old boss he'd greatly admired. The watch rarely left his wrist. Then he heard Beryl's quickened footsteps, the door opening, and Ted and Mona's welcome voices floating down through the air-exchange pipes. Jim turned off his lamp, got up from his desk, and ascended the stairs in his starched white linen shirt, jeans, and slippers, holding onto the railing; as he did, he caught a glimpse of Squirrelly's wedding gown hanging in a dress bag, next to the kids' old Halloween costumes. She'd splurged on it, spending eighty-eight dollars.

The years had been good to them. Jim had few regrets. His aging face reflected this peace, but just below the surface, two parts of him jousted, like old wrestling partners—inner contentment with a self-imposed rigor and intensity of spirit that only found satiety in the mind's busy work. As Jim emerged from his cave, he glanced back at his desk a moment. It had been a productive morning. Then he closed the basement door behind him. The sunshine on his skin would feel good, and he was looking forward to the company of old, dear friends.

* * *

"What's new, big guy?" Jim patted Ted's back affectionately, as they made their way, wineglasses in hand, out to the veranda.

"Driving up the Coquihalla this time of year sure is some stunning scenery. But it took us five and a half hours from Burnaby! There was quite an accident out of Abbotsford that caused a huge backup all the way up through Langley."

"Not surprising to me!" Jim leaned back in a deck chair and took a cool sip of Riesling, a crisp, white Okanagan wine with a distinctive floral and green apple aroma. It was one of Beryl's summer favorites.

He exhaled with silent satisfaction. Ted sniffed his glass and tasted the wine. "Ah, but well worth the trip," Jim's cousin-in-law exclaimed, with a boyish grin.

The girls emerged on the deck, already deeply engrossed in their catch-up. In the early years of raising kids and homemaking, the two couples had barely stayed in touch, only exchanging Christmas cards and letters or making the occasional phone call. But now that they had more leisure time, Mona and Beryl, Ted's first cousin, had become close, and as the social conveners in their marriages, the women made sure to book regular visits together.

"So Thursday, we've got reservations at Quail's Gate Winery, and there'll be live entertainment out in the vineyard," Beryl said, clapping her hands together.

Mona smiled excitedly. "I hear their pinots are incredible. Can't wait to try. Sounds great."

Beryl's eyes crinkled in delight. Then Mona looked out at the yard, admiring their manicured flowerbeds, vegetable garden, and giant peach tree, which had been recently trimmed and was starting to yield fruit.

"Looks gorgeous."

"With any luck, by the end of the week, we'll have juicy, ripe peaches to eat with ice cream."

"Mmm. I'll eat my first peach of the season from your garden."

Beryl's hand rested on her hip as she scanned the garden. It was coming along. Much of it they'd planted and tended themselves. But the last couple of years they'd brought in some hired help.

"The petunias are a little slow this year, but our vegetables are growing like mad. We just had our gardener here this morning to weed and cut the grass."

Despite the many material perks in life afforded by Jim's work, Beryl had never taken any of it for granted. If all she had in the world were Jim, her health, and their three girls and nine grandchildren, she'd consider herself just as lucky. Sighing contentedly, she threw her arm around Mona's shoulders, and the two squeezed each other tightly. With her daughters all grown-up and busy with their jobs and kids, it would be nice to have some female company around the house that week. The two fell into idle chitchat. What was being said was secondary to the joy each felt in the other's company.

"I like your blouse. It sure looks fantastic with those white capris. I need to pick myself up a pair."

Mona ran a hand over her pants as she sat down, crossing her tanned, fit legs. "Oh, thanks. I just got them on sale at an outlet mall in Burnaby."

"Oh, oh, oh—we've got a new plaza about five minutes from here. I haven't checked it out yet. Maybe we could pop in sometime this week?"

"Let's do it!" Mona beamed at Beryl. No distance or amount of time would ever change their friendship. Beryl had been there for them through everything, if not always in the flesh: all of the moments of joy and hope, laughter and tears of recent years. In that moment, Mona felt a surge of happiness in her throat and great love for her cousin-in-law, who, like her daughter, Anita, knew how to relish the simple pleasures in life.

Beryl slid into a chair, joining her friend. Despite their five-plus-hour road trip, Mona looked refreshed. That golf season she was playing eighteen holes four mornings a week with her ladies' league at the club. Beryl golfed too but not as often and wasn't nearly as competitive. More than anything, she liked to be out in the fresh air walking; so be it, if she was hitting a little white ball around along the way.

"Which courses do you want to play while you're here?" Mona pulled a Kelowna golfing brochure from her bag, and soon they were engrossed in planning their activities for the week.

By then, Ted and Jim had exhausted small talk and happily shifted gears to the familiar, comfortable territory of work and business. Both men were technically "semiretired," but neither took the title all that seriously. Each man ran his own consulting firms, sat on one or two key industry boards, and had his hands in a select few "boutique" projects. Jim had worked in the oil and gas industry for more than fifty years while Ted was a senior mining and metals guy. They were both professional engineers who'd graduated a decade apart from the University of Saskatchewan. The rings on their pinky fingers were a silent bond between them.

A disheveled newspaper sat on the table next to Ted. For years now, he'd scoured the business and world sections every morning for new developments, out of necessity but also with an insatiable thirst to know the context and culture in which he worked. No two days on the job were ever the same because no project ever developed inside a vacuum. Mergers, acquisitions, wars, politics, strikes, market shifts, new trade regulations, stock options, legal battles—all were part of a dynamic, shifting picture that could potentially influence how, where, when, and with whom they did business, or not.

Jim followed Ted's gaze over to the stack of papers, and then he tilted his head slightly as he recalled one of the day's headlines. "How about SNC-Lavalin?"

Ted shook his head. His face registered disgust and disbelief. "They sure are up shit creek without a paddle."

Jim nodded. "It's unreal, really."

Ted recapped the story. "Yeah. Duhaime's resigned, and investors are claiming damages of $1.5 billion."

Ted had worked with SNC-Lavalin's former CEO, Pierre Duhaime back in the mid-1990s, when the Montreal-based engineering and construction giant had purchased Kilborn, the employee-owned company that Ted worked for and was also a shareholder in. That new division was named Kilborn SNC-Lavalin.

Ted leaned forward in his seat, hands to thighs. He worked in a business that could sure play dirty. A class-action lawsuit had been launched earlier that spring by shareholders against SNC-Lavalin when its stock value dropped more than 20 percent. This move followed an investigation revealing more than $56 million in improper payments to the Gaddafi regime to secure contracts for infrastructure projects in Libya.

"What a fiasco," Jim remarked as he sipped his wine.

"If this was the US, Pierre would already be in jail." Ted shook his head. "Instead, he got a buyout package."

Duhaime had stepped down that winter as CEO of SNC-Lavalin when it came to light that he'd authorized the so-called facilitation payments, but not before his employer had agreed to pay him more than $5 million over two years. It had taken Ted aback when he'd learned that his former colleague was charged.

"He was an aggressive businessman, but I didn't think he was a crook."

Nodding, Jim leaned back in his chair and stretched his legs. The oil and gas business had its shady side, too; it seemed that wherever great fortunes stood to be made, scandal and blind ambition were not far off.

Jim scratched his head. "Did you sense anything...untoward...with those executives when you worked for them?"

Ted chuckled, glanced over his shoulder, and then looked Jim straight in the eye, from above his glasses. Over the years, through trial and error, he'd developed a sixth sense for double-edged dealers and their dirt. "You know, I was always impressed with the amount of work SNC-Lavalin could win, especially on projects in Quebec and the Middle East. I think some of that success was due to SNC-Lavalin being the largest engineering company with a French-speaking base."

In retrospect, Ted was grateful that, back in 1999, Duhaime got promoted instead of him. Back then, for two years already, Ted had held the title of president of Kilborn SNC-Lavalin, with Pierre running the division's Montreal office. Two months later, the Bre-X scandal went down. What amounted to the mining industry's own Watergate changed everything, from the face of the industry to Ted's own career path. Ted's employer, Kilborn SNC-Lavalin, had had the colossal misfortune of being the independent consulting company hired by Bre-X Minerals, Ltd., to assess the gold deposit at Busang in Indonesia. At the time, Ted was responsible for Kilborn's Global Mining and Metals Division. In May of 1997, he was sitting in an important two-day board meeting in Calgary attended by Bre-X prospectors and developers. Even then, it had struck Ted as strange that for such a key meeting, Bre-X's vice president, John Felderhof, was not present in the room; instead, he'd remoted in to the meeting from his place in the Cayman Islands.

During the Prospectors and Developers Conference (PDAC) one evening, Ted had joined his clients for drinks at the pub, where he found himself briefly shooting the breeze over a pint with Bre-X's exploration manager, a Filipino geologist by the name of Michael de Guzman. He was the guy responsible for "care and custody" of Bre-X's initial drill samples; the person who'd submitted the original estimate stating that the Busang site contained approximately two million ounces of gold, which sent stock prices soaring.

The world bought in heavily. And everyone in that conference room had stock options. So when the news hit the next day that Bre-X was a complete fraud, everyone felt the shock waves. It sent the Toronto Stock Exchange reeling. Investors had been duped out of billions of dollars. In the coming weeks and months, record numbers of people would leave the industry. Amid the scandal and allegations, there was a report, later that first day, that Guzman had committed suicide by jumping out of a helicopter somewhere in the middle of the Pacific Ocean. A body was never found, and many would speculate over whether he'd faked his death and escaped or was murdered.

Because Ted was in charge of the global division at the time, he had faced consequences. He wasn't personally responsible for the failure in identifying the tainted drill samples; it was an engineer general manager on his staff, by the name of Paul Semple, whose relative inexperience had likely led him to overlook suspect uniformities in the ore body—although even a skilled eye could have missed them. Ted had been disappointed that Paul hadn't caught it. But the guy wasn't guilty of anything, except, maybe, optimism. He hadn't done anything illegal or immoral. So everyone looked to Ted for answers. He provided the necessary leadership, and then the legal department took over.

The facts soon came out: Bre-X had already falsified their drill samples long before any of Kilborn's engineers got involved. Kilborn Engineering was cleared of any wrongdoing; still Ted ended up carrying the can for his team. After Bre-X, the "Kilborn" name was dropped from the company's moniker, and Ted was overlooked on the major promotion. Despite the fact that Kilborn executives and field workers had been cleared of any wrongdoing in the elaborate gold-tampering scheme, the mere, if tenuous, association with the crooked Calgary-based company ruined Ted's chances of moving up at SNC-Lavalin.

"I was tarred with the Bre-X brush. There was nothing I could do about it. My boss, at that point by the name of Taro Alepian, was trying to reorganize Mining and Metals worldwide, and it became quite obvious to me that he couldn't sell my name to the board."

Jim sat, folded his arms, shaking his head. Ted paused to take a sip of his wine, reflecting a moment. Soon after, Ted was sent to Siberia on another project. Pierre continued his climb to the top of the SNC-Lavalin ladder, until, apparently, there was nowhere for him to go but down.

"I wonder what the heck was going through Pierre's mind when he'd signed his name on that dotted line," Ted puzzled.

"Whatever the case," Jim cleared his throat, "he signed."

Ted nodded. "That's the bottom line, isn't it?"

Jim nodded, silently.

"Apparently even Pierre's chief financial officer had refused to ink the deal because it smelled of a bribe."

At any rate, Ted was glad he was not in Pierre's shoes.

"If anyone offered me a significant anything to grease my palms, they'd be off the bidder's list, plain and simple," Ted spoke emphatically. "Even if you don't accept the bribe, it's there. You just don't want to work with those people."

"No sirree." Jim swirled his drink in his glass, paused a moment to glance at their wives, who were deep in conversation, and then he turned back to Ted. He could chat like this all afternoon. These were topics he could really sink his teeth into. Already, he felt twenty years younger than he had that morning.

Jim summed up his view. "This Pierre didn't make the right call. And now he's paying for it."

The fallout of such decisions might not bite a person in the butt right away, but it would bite them eventually.

Ted took a sip of his drink and then ran a hand through his hair. "I don't think he'll get a job in the industry again."

He crossed an ankle over his leg. Years in this business hadn't turned him into a cynic; they'd made him shrewd and uncompromising, if a little bullheaded, at times. He'd rather stand on his principles than be potentially taken down by someone else's flimsier ethics.

Jim hunched over, legs spread, examining the deck. For a moment, both men were silent. Both of them had made it to where they were that day by doing their damnedest to make the best calls, as they saw fit. Around the world there were many ways of conducting business. In the end, all you had was your reputation and whichever moral compass you chose to follow. Over the years, Ted had had suppliers who'd taken him to hockey games in the spirit of relationship building, but he'd made sure never to cross that fine line for a favor.

Without another word on the subject, Jim and Ted raised their glasses.

Ted let out a satisfied sigh. "Hey, did I tell you about that time when I was in Saudi?"

Jim rubbed his nose thoughtfully. "I'm not sure. Let's hear it."

Ted dove into his tale, recalling the meeting in a London hotel room back in 1997.

"It was a one-hundred-million-dollar lump-sum turn-key job." Ted was negotiating a deal for Kilborn SNC-Lavalin on a small Saudi Arabian gold mine, pre-Bre-X. Jim inched his chair forward, relishing his cousin-in-law's every word.

"We'd sent in a competitive bid, and the Saudis had accepted it. So I flew out with my projects and technical managers, and in the course of our initial testing, we discovered inaccuracies in their process-flow sheet."

The two men were speaking the same language. Their wives glanced over amused, watching their two relatively antisocial husbands gabbing like schoolgirls, a rare sight. Ted was talking with his hands, and Jim was perched on the edge of his seat. Mona and Beryl shook their heads, eyes shining silently, and then returned to their girl talk. The sun was shining, and the cool white wine was going down easily. All in all, it was shaping up to be a lovely afternoon.

"You see...they'd based their proposal on faulty information provided by the Saudis. Now, their client was telling them that any additional expenses were at their own cost. We went back and forth, but couldn't reach an agreement." Ted spoke in his boardroom voice. Finally, the son of the minister of mines asked to meet Ted in London. His client invited him up to his hotel room.

"That's when he said, 'If you give me one million dollars, your problem goes away.'"

Jim leaned in, excitedly. "Good lord!"

"I tell him I need to think about this. So I call my boss, Taro Alepian. I tell him, 'He's bribing me.' Taro says, 'You do whatever you think is best.' He's not saying no, but if something happens, he's really saying, 'It's not my responsibility.' I was shocked."

"So what did you do?"

The back of Jim's calf had fallen asleep. For a moment, he ignored his temperamental leg. It wasn't every day he got to sit around and share old battle stories. He loved it when his cousin-in-law came for a visit.

"Well, someone wanted to be paid. I decide I'm not going to do it. I tell him, 'It's against your laws. It's against our laws.' I know it's against my principles. They'll lop your hand off or something over there. But that's not even the point. There's no stopping the effects of such a decision. It's irreversible."

"So what happened?"

"The deal died. I walked away. Gold prices had gone down, and the Saudis weren't keen on the project anymore."

Just then, Beryl cut in. "Sounds like conversation is getting a little heavy over there, guys," she teased. Actually, she was thrilled to see her husband enjoying himself. She could see his brain cogitating, and hopefully, she thought to herself, amused, it was getting his heart rate up a little. "How about a toast?"

Ted grinned at his cousin. "Great idea!"

"To the start of an awesome week with you guys." Beryl raised her glass, and the four clinked their crystal together.

Just then, there was a loud crack, and Jim nearly jumped out of his seat. "What the...?"

The others were unfazed by the distant noise and surprised by Jim's reaction. But it sounded much, much louder through the sensitive filter of Jim's wakeful mind. Although it was only an engine backfiring on the street next to them, the sound dislodged a memory in Jim's psyche. He looked at his Squirrelly, who knew all too well what was happening. She laid a hand on his shoulder and expertly diverted the conversation to dinner plans for that night, until her husband could regain his composure.

For a moment, all Jim could hear was gunfire in Turkey reverberating inside his brain, like a ghost trapped inside a maze, stuck haunting for an eternity.

> **LESSONS LEARNED**
> 1. Always maintain honesty and integrity. One slip can ruin your reputation forever.
> 2. The senior manager is always responsible for the actions of his people.
> 3. On all major projects there is always a mix of business and politics. You must not compromise the business principles in favor of politics.

CHAPTER FIVE

Dodging Bullets

The drilling activities were located in southeast Turkey in a remote and unexplored valley that some say was guarded by gods. From dawn to dusk, the slanted shadows of oil derricks creep across the ancient landscape, sleepless specters of modern-day enterprise.

This land is rich in history, dotted by the timeworn remains of mosques, churches, castles, caves, and tombs. Once, a megalomaniac king built for himself a lavish burial site atop Mt. Nemrut, dedicating it to the Greek deities. In grand fashion, he had the giant heads of Apollo, Zeus, Hercules, and himself, King Antiochus I, carved into the rock for all future generations to see. But his immortality plan failed, as this storied land is also rich in coveted gold, petroleum, iron, coal, and fertile soil. From the great banks of the Tigris and Euphrates Rivers, and since the dawn of civilization herself, wars have been waged in the name of its abundant resources and real estate. A century later, the illustrious kingdom was sacked by Romans, and its sacred sanctuary destroyed, the regal idolatry reduced to rubble in an attempt to erase the Greeks' former glory from time's record.

If these mountains could speak, they'd recount the rise and fall of many empires and the confluence and diaspora of millions of people, whose lives and livelihoods could change, like the wind, ever subject to the elements of kingly decree. No history is carved in stone. Yet, supposedly at night, if you listen carefully to the howling mountain winds, you'll hear the stories of forgotten refugees: a Greek, a Roman, an Arab, a Jew, a lost shepherd in search of his flock, or a banished soul ever seeking a place to rest his weary head.

Today, it's the displaced Kurds, whose fate on these lands is still to be determined. An ancient nomadic people of mostly Sunni Muslim faith, they've long lived in the rugged region that now shares modern borders with Turkey, Iraq, Iran, Syria, and Armenia. Today, fifteen- to twenty-million Kurds straddle state lines and make up the largest ethnic group in the world without a nation-state to call its own. When the Ottoman Empire was carved up after World War I and divided between the Turks, Arabs, French, and English, the Kurds quite simply got stiffed in the deal. They've been gypsy outsiders in their native homeland ever since. As in that absurd child's game of Musical Chairs, the exiled Kurd, like a modern Israeli or Palestinian refugee, is the poor kid left standing without a seat when the music suddenly stops. There's a saying, in these parts, that "Kurds have no friends but the mountains."

This is a story of how friendship survives, despite vast differences because it can move mountains.

***

Jim sat at a table inside his trailer hunched over a pad of paper. There was a mechanical pencil tucked behind his left ear and two more poking out from his pocket protector. His lips moved faintly, but he was otherwise still. His far-away gaze was like the intense blue Aegean sky. Five minutes went by. Jim hadn't budged. He was lost in a complex problem, trying to solve an emergent drilling issue that had arisen in a geological formation they'd encountered in the well. This could put the corporation and the Turkish team behind schedule.

He'd learned, time and time again, that not all projects go exactly to plan. External factors outside your control could always affect your progress. But like all good managers, Jim knew the importance of contingency plans. He'd make the necessary adjustments. In his eyes, this philosophy distinguished the Audis and BMWs of the business world from the Yugos and the Ford Pintos. In the long run, it paid to err on the side of sound thinking. Back at home, he'd recently told his grandson, Zac, who was saving up his allowance for a new bike: "Always buy the best you can afford." Quality was important, but quality in projects meant meeting specifications not adding unnecessary and costly extras. Jim was an unsentimental realist, as down-to-earth as they come. But everyone has blind spots. Sometimes you just can't see life's little reality checks coming.

Jim leaned back easily in his chair, reaching for a sip of coffee. He wondered how the corporation was fairing on the TSE in light of ongoing Kurdish rebel uprisings against the Turkish government. No news was good news when you were living out of a trailer in the middle of the mountains. Southeast Turkey also just so happened to be a war zone. Jim checked his watch. Here, in these parts, time was as blurry as its national borders.

Something suddenly moved in Jim's periphery. He lifted his eyes from the desk to see a small boy approaching his trailer. It was early that Sunday morning, and Jim could also see that the boy's mom and dad had accompanied him. They'd stopped some distance short of Jim's trailer but encouraged their son to go on. The boy hesitated, but Jim opened the trailer door to invite the boy to visit. Gaining confidence, the boy walked right up to the trailer door. Jim estimated his age to be seven or eight, about Zac's age.

"Well, hello there," Jim said, kindly, to the barefoot lad, who was clutching a plate of food in his hands. "What is your name?"

The boy remained silent but shuffled on the spot, and then glanced back over his shoulder. Jim followed his gaze to the boy's parents, who were dressed in typical Turkish clothing, watching and waiting some distance from the trailer.

Jim nodded his head, and the boy's dad, who was tall and steady, returned the gesture. His wife, who had an infant tightly bound to her robed body, smiled warmly.

"My name is Jim," he said, pressing his palm to his chest. He pointed in the boy's direction. "What is your name?"

"Diyari."

"Diyari. It's very nice to meet you. And what have you got there?" Jim asked, motioning toward the plate in the boy's hands.

He handed Jim a dish of what looked like rolled crêpes soaked in honey. Surprised and pleased by the sweet scent hitting his nose, Jim's eyes lit up. Diyari grinned. Jim picked up a sticky crêpe with his hands, and the wide-eyed child watched him intently as he took a bite.

"Mmm." No translation necessary.

"Did you make this?" Jim asked, lightly touching his index finger to Diyari's small chest.

The boy twisted around, pointing excitedly toward the low mountain range in the south and spoke rapidly in a foreign tongue. "Hingiv! Hingiv!" he sang proudly.

Jim gestured to the food on his plate. "Hingiv?"

The boy shook his head.

"Hingiv?" Jim asked again, this time sticking one finger comically into the honey and holding it up to let the sticky liquid drip slowly down his hand.

The boy laughed and nodded. "Hingiv!"

"Hingiv is honey!?" Jim exclaimed, and then the teacher in him sounded out the English for the boy. "Honey."

"Honey," Diyari repeated back.

Jim pointed back south toward Syria and the expansive Taurus mountain range. "Do you live up there, in the mountains? Is that your...home?" he asked.

"Hom?" said the boy tentatively, confusion in his eyes.

They were a family of herdsmen who moved their sheep higher in the mountains in the summer where the temperature was cool and there was plenty of food and running water. They also tended a number of beehives, from which Jim was given that very thoughtful gift. Diyari liked the funny white man but didn't understand his strange talk. So he shrugged his shoulders and idly dipped his small index finger into a pool of honey that had dripped onto the table.

"Hingiv," he said, smiling widely and then turned to leave.

"Wait!" Jim suddenly jumped up to grab something he remembered in his briefcase. A few seconds later, he pulled a small Canadian flag pin out of a Ziploc bag and handed it to the boy. The boy's eyes lit up. He took the gift, cradled it carefully in his palm, and then ran off.

"Thank you for the...hingiv!" Jim called after him.

The boy laughed, and then sticking his fists into his armpits buzzed like a bee, stirring up dust, as he ran back to his family.

Jim instinctively glanced at his watch. It was 1:00 a.m. in Calgary. His own family was still sound asleep. He imagined them waking up in a few hours, the smell of flapjacks cooking on the stove and Beryl out in the garden picking berries and vegetables for the day, the dewy folds of her pink English roses opening to the sun. He sighed. Just a couple more weeks and he'd be home again. When he got back, he decided he'd invite the entire family over for a big Sunday brunch of crêpes and honey.

"Hingiv," he said, shaking his head in amusement, and then he returned to his desk.

The next day, Diyari's dad came back to Jim's trailer to retrieve the plate.

* * *

If those great mountains could perceive all, right that moment they'd hear Jim say, "Good Lord, what have we got going here?"

The first gunshots were fired at around 11:00 a.m. At the time, Jim was sitting inside his trailer, which served as his sleep quarters and office. He was wrapped up in reviewing drill data from the previous day and preparing a report. When he heard gunfire, he jumped up from the table.

"What the...?"

He peered out the large window above his desk from which he could see the entire operation. He had a clear view up the catwalk to the derrick and into the mountains beyond. A skirmish was underway in a saddle-shaped area at the crest of the mountain. As Jim stood there, eyes glued, he saw one militiaman jump for cover. Jim instinctively ducked and reached for his transmitter to contact the drilling contractor's supervisor, Bob.

"It's them, isn't it? The rebels? They're here."

"Yes, they're here. The militia is on it. They're fully capable of handling the situation." They heard another loud crack in the distance.

Months ago Jim had been briefed and prepared for the possibility of such an attack, and he'd filed this information away inside the cabinets of his logical mind, focusing on the task at hand—until now. If he hadn't heard the rumors of other rigs being attacked, he probably would have put the gunfire down to farmers shooting rabbits. But he knew there was a very real possibility that this might happen since the Kurds had recently visited a nearby camp, lined up some of the people under a lamp standard, including the on-site engineer, and taken them out. Jim pushed the thought from his mind and focused on Bob's voice in the receiver.

"They're way up in the hills. There's no immediate threat to us. Let's just continue going about our daily business." Bob spoke rationally.

"We'll continue drilling," Jim affirmed. Contingencies were made for moments like these. The situation was under control. "Let's go for lunch."

By the time they were sitting down to eat, it was early afternoon. Jim could barely taste the roasted chicken and beans he'd piled on his plate but swallowed bite after bite as fuel for the uncertain hours ahead. As he and Bob were finishing up, more shots rang out.

"Sounds like bazookas," said the supervisor. "They're closer too."

"I think we'd better let them know in Ankara," Jim replied. "It's time." Head office could contact the army, which could provide backup in the event that the situation deteriorated. This was starting to appear more and more likely. Together, they made the call.

Jim headed back to his trailer and Bob, the committed individual that he was, returned to the rig floor to continue drilling. Over the next few hours, Jim worked on edge in his office.

He was concerned about the militia's ability to provide their operation with adequate security. The mountains may be more like large hills, but they were rocky. The militiamen carried rifles but wore only sandals on their feet and cumbersome Turkish robes, which draped around their bodies and served as headgear. Could these men defend them if push came to shove and the rebels gained critical ground in the next few hours?

Jim caught himself staring at his transmitter and thought about calling his wife in Calgary. Their only communication in and out of the site was by satellite. Costs were high, somewhere around three dollars per minute, so phoning home was done infrequently. Then Jim pushed the idea from his head. He would not call Beryl. He'd save her the worry, which, he told himself, would probably end up being over nothing at all. Then he turned his attention back to his report and tried to ignore the distant gunfire. He wasn't so sure. Mind over matter, he told himself.

By late afternoon, the face-off between militia and Kurdish rebels had begun to converge on the perimeter of their operations. Ken, the site geologist, showed up at Jim's trailer door. He'd been in the mud-logging unit with the Romanian workers when they finally decided it was too risky to continue their work.

"She's getting closer and more frequent. We packed it in. Told the group we're gonna meet by the pipe rack in five minutes to discuss our safety plan."

Jim nodded, pushing his work aside, and stood up. Out the door, he could see by the smoke trails that a lot of tracer bullets were flying about. This was definitely not a farmer shooting rabbits, Jim thought. He took a deep breath. He was ready; for what, time would tell.

A few moments later, Jim joined Ken and the other rig personnel near the rows of pipe racks in front of Jim's trailer. When the first bullets started deflecting off the derrick, Bob finally gave up drilling and made haste to join the rest of the crew. They all gathered by the water tank. On the advice of a security expert who'd visited the rig earlier, they'd chosen this spot for its safety in the event of gunfire. The place was set into an excavation next to the hillside. If the rebels got too close, all they could do was take cover under the water tank and wait for the army to arrive. Jim checked his watch.

"It shouldn't be much longer now," Bob said, reading his mind.

Suddenly their contingency plan felt flimsy. Take cover and wait for the army to rectify the situation. Jim wasn't good at sitting back and waiting for someone else to solve the problem. But he had no choice in the matter.

Days earlier, they'd all gotten the lowdown by government officials on the intensifying armed conflict between the republic of Turkey and the Kurdistan Workers' Party (PKK), a radical militant group of nationalists fighting for independence in their ethnic homeland. Recent air strikes and ground incursions by Turkish militia in the southeast were a response to escalating guerrilla violence both at home and abroad. Government forces were reportedly targeting strategic PKK hideouts and bases and in some cases evacuating and destroying entire Kurdish villages. Spotty state-radio news clips reiterated what was happening on the other side of their compound. The military campaign and uprisings were turning the Turkish countryside into one big refugee camp. And the oil well was smack dab in the middle of it. They had to act as a team, stick together and keep safe until the army resolved the situation.

Jim resolved to stay calm. After all, getting worked up wouldn't help. This was, quite simply, the collateral risk of doing business in a part of the world plagued by civil unrest. They knew this could occur; now it was actually happening. So they just had to deal with it. Jim ran a hand over his head. Then he glanced at the kid with the big machine gun, a member of the Turkish militia who was guarding their group. The kid couldn't be a day over sixteen, Jim thought. He looked awkward holding his weapon and also extremely nervous, which did nothing to inspire Jim's faith in their security. He reminded himself that he wouldn't even be there if it weren't for vigorous guarantees that, in the event of such an uprising, armed militia would protect the camp and all its personnel.

Last year the PKK had attacked Turkish consulates and commercial facilities in a number of western European cities. The Turkish government apparently wasn't taking any chances on a high-profile project like theirs. For weeks now, they'd been guarded around the clock like a corral of sheep. Thirty militia with AK-47s watched over them, keeping them safe from the wolves in the hills. In the beginning, the militia's presence was unnerving; possibly, even, unnecessary, Jim had thought. Now he was just grateful they were there to defend them.

The sun was slowly disappearing behind the mountains. Soon they'd be cloaked in darkness. Whether he and his crewmates liked it or not, they were in the middle of it. Jim hoped their young soldier had shown up more than once for target practice. For the first time, he wondered if he'd been out of his mind to travel to this country in the first place.

Suddenly, Bob jumped up from the milk crate he'd been sitting on. "Hey, where is Gerritt?"

Everyone looked at each other blankly. The Dutch driller was nowhere in sight. Somehow he'd gotten split off from the group and never made it to their muster point. Jim was about to go and look for him when there was a loud crack nearby.

Their young soldier opened fire, letting a spray of bullets rip at the hillside.

"Stay low, out of sight! Allah kahretsin!" he yelled at them, eyes wide. Jim had never seen such terror in a young man's eyes.

He pointed to the water tank. "Go! Now! The rebels are close!" Jim and his team moved quickly, heading for cover. Wherever his colleague was, he hoped Gerritt had found a safe hiding place.

By nightfall, 110 Turkish army troops arrived at the well site in personnel vehicles with listening devices and camo paint. They knocked out the lights and cut the generators. The camp was engulfed by darkness.

Two lieutenants, both named Omar, led the soldiers. One was chubby, Jim noted, and the other one looked fit with a slight athletic build. The chubby one stayed with the crew and sat down next to Jim by the water tank, rifle between his knees. The other Omar left, taking three army personnel with him into the mountains.

Jim turned toward the waiting lieutenant, eyes questioning. Omar answered calmly, in clear English. "They will go into the hills now. Take care of the situation."

Jim nodded, saying nothing. He was relieved to note that their army dress was far more suitable for the terrain than that worn by the militia. They had lightweight boots that looked strong and flexible, which would allow them to move quickly over the rocks. These thoughts occupied Jim's mind as they waited.

Then Gerritt suddenly appeared. Everyone shuddered, and Omar raised his gun.

"Don't worry, it's just me," came a loud whisper.

"Thank, God." The others sighed with relief.

It was dark, but as he peered up he could make out his colleague's large frame.

He nodded at Omar, who lowered his gun. "It's him."

As Gerritt approached, they could see he had something tucked under one arm that looked like a bottle. Alcohol was strictly prohibited on-site.

"What have you got there?" Jim asked.

Gerritt raised the bottle and simply said, "I felt under the circumstances, this might be an exception to the rule. I got caught in the loo."

A few of the guys laughed, and then others joined in. They were just glad to see their mate OK.

Gerritt raised the bottle.

"Skål."

"Skål."

***

For hours, they waited. The lieutenant offered Jim a Kalashnikov, but he shook his head and instead made sure his passport was handy. He searched for the handful of Canadian flag pins in his pocket and squeezed them in his palm. That's when Jim thought about gods on top of mountains and the little Kurdish boy, no older than his grandson, who delivered him crêpes and honey on Sundays. Come what may, he would plead innocent. That was his plan; it was not, and would never be, to point a Kalashnikov at any human being. Besides, Jim was inexperienced in the use of army weapons, and he knew that if one of the Kurds approached him with a gun, there would be no contest. If he were to be lined up under a lamp standard like those unfortunate personnel, Jim knew his best defense was proof of his Canadian identity. The comfort he drew from knowing his own resolve got him through the next several hours. A calm washed over him that came from a deep inner surrender to a situation unfolding that was clearly beyond his control.

The gunfire moved back into the hills. Everyone started to relax a little and buckled down for the wait. Jim and Omar became instant friends, falling into lively conversation about their jobs and their families. But each man knew the friendship would be short-lived; they were just passing time, making the most of the situation, until the uprising was over and the convoy could depart.

In the back of his mind, Jim wondered whether, if faced with the barrel of a gun, his alibi of nonparticipation would be enough. Could a small red-and-white souvenir pin save his life? He couldn't know. It was true: this was not their battle, but whether "the enemy" would see him as an innocent bystander or a disposable decoy was unknown. In the darkness and relative seclusion of their hiding, Jim yielded to chance. For the first time, Jim felt like an intruder in someone else's conflict. Business was business, but what business did they have, really, with people who negotiated so cold-bloodedly? Were the human risks of elite partnership with a warring host worth their weight in black gold?

By midnight, the standoff was over. The lights were restored. They were still waiting in their hiding place.

"It's clear," a voice finally said in a thick Turkish accent.

As Jim and crew members filed out of their hiding place, Jim bowed to the young soldier who'd fought in the hills for them that night. He placed his hand over his heart.

"Thank you," Jim said. "We are grateful for your help."

The soldier just nodded. He looked now much older.

***

The crew went back to drilling. An eerie silence diffused the valley like the smoke of phantom gunfire, still ricocheting inside Jim's head. As he sat inside his trailer by the window, the wounded militiamen started descending from the hills. The nearest hospital was an hour away, and so he pushed his work aside, opened his first-aid kit, and did what he could to help patch up the injured before they left. By that point, as Jim saw it, this was just part of the job.

When the last wounded soldier left in the convoy, the sun was coming up. The crew got together to discuss the last twenty-four-hours' events and what they meant to rig operations. They concluded, sensibly, that this was just a hiccup.

Before heading to sleep for a few hours, Jim updated his report and made best projections for the short-term future. Time lines would likely be affected, but cost factors and project perimeters could be revised to minimize change and cost overruns. It was business as usual. After all, hiccups happen.

***

Several militia and members of the Turkish army would stay on-site for the next several days. It would turn out that there would be nothing much for them to do except pester crew personnel for AA batteries to use in their night-vision goggles, which they'd been outfitted with by the army, without disposable power cells.

Ten days later, while listening to local police radio through an interpreter, Jim and his team learned that following the attack, the Turkish army had gone after the Kurdish group and killed all twenty-two, including two women. Jim lowered his eyes and just shook his head. Everyone was disturbed and troubled by the news. Jim was relieved when the meeting ended and he could retreat to his trailer. As he sat at his desk, he idly wondered if tomorrow he might see his little friend, Diyari. He didn't count on it.

The boy hadn't stopped by last week—probably because his family had also suffered from the visit by the same Kurdish bunch, only to have food and other supplies stolen from them during the Kurds' retreat. Jim noted that more than thirty thousand Kurdish people have been killed thus far in this ongoing war for autonomy, which they'd gotten mixed up in.

The European Union and NATO confirmed that while some Kurdish villages had been destroyed or evacuated, others were converted into pro-government villages and their local shepherds and farmers recruited to the village guard to stop PKK from operating in their communities. This was reportedly driving Kurdish rebels into the mountains, where they were launching retaliations on any pro-government settlements, including assaults on regular civilians.

Nothing was ever black and white. Wherever Diyari and his family were, Jim hoped for their safety and protection. He also prayed that this war might end, sooner than later.

"Just a hiccup." Jim tapped pen to paper, not totally convinced. Then under his breath, he slowly sounded out the foreign words, which seemed bittersweet on his tongue: "Hingiv."

***

When the drilling operation was complete, before Jim left the mountains for home, he gave away his work clothes to the Turkish crew. They were grateful, as some could not even afford socks to wear inside their work boots. On one previous occasion, Jim had given one of the crew a pair of socks, and the man had tucked them under his shirt and taken a back route to his own trailer so that his good fortune wasn't noticed by the others.

As farewell gifts, the crew gave Jim a string of authentic Turkish beads and a pair of baggy pants with the crotch somewhere around his knees. They were very proud of their offerings, and Jim was grateful for their thoughtful gifts. He still has the beads but is currently unsure of the whereabouts of the pants.

> **LESSONS LEARNED**
> 1. Quality is important but quality in projects means meeting specifications not adding unnecessary and costly extras.
> 2. External factors outside your control can affect your progress. Contingency plans are important.
> 3. Risks and benefits need to be understood before conducting business in a part of the world plagued by civil unrest. The country must be sufficiently stable economically and politically for long-term investment.
> 4. You have to be sensitive to the culture of the local people.

CHAPTER SIX

Memories

"Any day now, Jim!" Beryl chided her husband.

Ted was tapping a domino between his finger and thumb against the patio table. He was anxious to make his next move. Mona laughed. Her cheeks were aglow with the Cooke's own special reserve merlot that Beryl and Jim had bottled that winter.

"I'm stuck with all these threes! Ted, can I trade you?"

"Nope!" Ted grinned.

The foursome were into their second bottle of red wine and had switched over to home brew. They were getting a little rambunctious over a round of Mexican Train Dominoes, one of Jim's old favorites. The object of the game was to be the first player to get rid of all of your dominoes by placing them on the ends of trains extending out of a central hub or "station." It had been several minutes since the last move.

"He's deliberating," Ted said knowingly, eying his opponent.

"Well, he's on Mexican time, apparently," Mona teased, sending Beryl into a small fit of laughter.

"He thinks, and he plans, and he strategizes—and everyone waits!"

"It's payback, Beryl." Ted winked at his cousin.

Finally, Jim snapped his domino into place, stealing Ted's spot.

"Oh, you didn't!" Ted smacked the table in mock frustration.

Everyone laughed. Then an old, popular Leonard Cohen song drifted through the warm night air, from wireless speakers on the nearby table. "First we take Manhattan, then we take Berlin."

"I love this song!" Beryl gushed, suddenly remembering something she wanted to ask their houseguests. "Ohhh, by the way, what are you guys doing October 5? We've got a couple of extra tickets for Leonard Cohen in Victoria at the Memorial Centre. Can you join us?"

Mona and Ted looked at each other, hesitating for just a split second. October 5. That date would forever haunt Ted. Just like the random sound of a backfiring engine could jolt Jim, this fateful day so many years ago had made Ted a different man.

"Are we around then, hon?"

"I think so."

Then sitting up a little taller, Mona turned back to Beryl and Jim, sounding upbeat. "We'll check. That sounds like fun. I've never seen him in concert before. You could stay with us in Burnaby for the week."

Beryl clapped her hands excitedly, and Ted smiled warmly at his friends. A crease had worked its way across his forehead. By this stage in the game, he hid his distraction fairly well, but in the back of his brain an old reel started playing. Not a day went by that he didn't think of Anita or Kumtor.

CHAPTER SEVEN

Kyrgyzstan Assignment

It was July of 1995. Ted was on his way back to Central Asia to develop the Kumtor Gold Mine. Two weeks before his boss had rung him up at the new Toronto office.

"We need you, Bassett!" John Somers had said definitively.

"Where do you need me?" Ted was caught off guard. He and Mona had just settled into a two-story house in the quaint, lakefront township of Oakville. Ted was commuting daily into the city, where he worked at Kilborn's head office as senior vice president and general manager of Eastern Canadian Operations. He'd just gotten a diamond membership at the Oakville Executive Golf Club and was enjoying the warm, easy days of an Ontario summer. When he heard Somers' next word, his shoulders crept toward his ears.

"Kyrgyzstan."

"What? Isn't Larry handling things over there?"

"He quit. I don't blame him."

"He didn't last four months! Was it the lifestyle or working for Mike?"

"Both."

In September, Larry Chandler had been nominated to replace John O'Hagan, who'd walked off the job, citing irreconcilable differences with their Cameco contractor, Mike Lindberg; by Mike's account, John was "unacceptable" to

Kumtor. Ted had done the feasibility study and was the senior client representative for the Kumtor project, so he was asked to take over until a suitable replacement could be found. There was no question that Somers needed him.

That night Ted talked with Mona. After working out a contract with Cameco, within weeks he was on an airplane headed back to the old Soviet state. Ted thought he knew what he was up against. As it would turn out, Mike would be the least of anyone's problems.

***

More than two years had passed since Ted's initial visit to Balykchy, Kyrgyzstan, during which he and his Kilborn-Cameco team had collected assay samples with the help of their new Kyrgyz friend, Altin Jyrgal Uulu, and his Russian wingman, Vlad. All appeared as good as gold, but within a matter of months, a scandal had tainted the multibillion-dollar project.

A shady commodities trader by the name of Boris Birshtein, with suspected ties to the KGB, allegedly smuggled more than a ton of state-owned gold out of Kyrgyzstan in a private helicopter to a Swiss bank. The founder of the Toronto-and-Zurich-based company, Seabeco, Birshtein also held a seventh-floor office in the White House and was personal advisor to first Kyrgyz president, Askar Akayev. Following an internal investigation, Birshtein stood accused of embezzling gold in collusion with an inner circle of high government officials.

President Akayev—an engineer and physicist, schooled in Russia, engaged in academia, and turned politician overnight—survived the congressional scrutiny. However, his vice president, Feliks Kulov, was forced to resign amid cries of corruption and his prime minister, Tursunbek Chyngyshev, left office after a nonconfidence vote in parliament. Russian intelligence agencies widely suspected Seabeco as a front for financing a Communist Party in exile. Furthermore, it was Seabeco's founder, Birshtein, who'd paved the way toward a lucrative mining partnership between Kilborn's client, Cameco Corp. and the country of Kyrgyzstan.

The deal soon fell under public scrutiny. Why had the Canadian mining company—and world's largest uranium producer—been awarded rights to the Kumtor gold deposit, ostensibly, without any competition from rival bidders? In light of the Seabeco scandal, parliament raised the question, and many fingers pointed to Birshtein, their acting agent, who'd unilaterally pursued Cameco for engineering, construction, and development services.

By 1993, the fledgling Kyrgyz republic's constitution already included a solid framework of antitrust laws to protect against unfair competition and monopolies. Just as with marriage vows and professional oaths, the law was a set of governing rules. People ultimately upheld or broke them. No project could manage the variables of human error and impropriety entirely. Gold may glitter and blind; greed and greatness were two sides of the same coin.

Ted's client defended the Kumtor contract as legal, both under Kyrgyz and Canadian laws, but stated publicly it would "think twice" about engaging Birshtein as an agent in the future; the company, however, noted, he'd done a fine job brokering the deal for them. Now Cameco was eager to seal it—and gain access to a tantalizing piece of the international gold market. The Canadian company had already invested $6 million in the Kumtor project, despite Kyrgyzstan's well-known track record for double-dealing.

From Ted's—and Kilborn's—perspective, as Cameco's engineering guys, they'd never seen any corruption and no one had attempted to pay them off. Clean hands and due diligence were their modus operandi, and so they moved forward.

Then, just four months before Ted moved to Bishkek in July of 1995, the first Kyrgyz government fell in a civil uprising. Amid accusations of endemic corruption, President Akayev and his family fled to Russia. He'd return and be reelected later that same year, with 70 percent of the vote and fraud heavily suspected. By 2010, Akayev would be back in Russia, facing extradition charges for stealing billions from the state budget, usurping power, killing civilians, and various other crimes.

But none of this would stop the Kumtor project from proceeding. By the summer of 2012, as Ted was kicking back on a Kelowna porch with his cousin-in-law, the Kumtor mine would have produced more than eight million ounces of gold. The coveted mineral would account for nearly half of the country's exports, yet Kyrgyzstan would still rank the second-poorest country in Central Asia.

***

But it was still the summer of 1995. Ted and his team were about to learn about the opportunity costs of some projects; they could be extremely pricey and dicey. In some cases, no amount of gold could be unearthed to ever repay them.

> **LESSONS LEARNED**
> Don't be afraid to take on challenging assignments regardless of the inconvenience to your life. Challenges should be treated as opportunities. They also enhance your professional and personal development. You don't know what you're capable of until you test the waters.

CHAPTER EIGHT

On the Way

The odds of dying in a plane crash are about one in eleven million. Statistically, your chances of fatality increase
    1. during takeoff and landing;
    2. when seated in the rear;
    3. while airborne during a storm;
    4. en route to South America or Africa;
    5. when mechanical problems arise; or
    6. if a pilot breaches protocol resulting in flight error.

Then your odds go up exponentially—to about one in twenty thousand. Freak death by lightning strike is more likely.

But Ted wasn't pondering probabilities as he stretched out his legs in business class. A scientist at heart, he trusted in the superior design and engineering of the Virgin Atlantic 737 and wouldn't lose sleep on the nineteen-hour-long haul from Toronto to Kyrgyzstan via Heathrow. In fact, finally with a few hours on his hands, Ted was eager to dig into the New York Times best-seller Mona had packed in his carry-on.

He took a sip of white rum and Diet Coke, adjusted his head pillow, and then cracked open a crisp, new copy of the late Harvey Penick's book *And If You Play Golf, You're My Friend*. The great golf instructor had died earlier that spring at the age of ninety. Ted was determined to take three strokes off his game that year, even if he'd be spending part of it stationed at a remote open-pit mine, four thousand meters above sea level inside the treacherous Tien Shan Mountains and a four-day drive to the closest tee-off in neighboring China. With any luck, the Kumtor project would improve his handicap. An avid sportsman, Ted liked to joke with colleagues that his golf game was his best barometer for how a project was going.

"There's nothing like crystal-clear scope, on-track scheduling, cost containment, human competency, and proven project deliverables to rock my short game and put the magic in my back swing," he'd say, as he nursed a martini and discreetly checked his wristwatch at some obligatory cocktail party.

The more Ted planned, the luckier he got. But as every sane and sensible project manager knew, that fickle thing called "reality" hijacked even the most brilliant plans eventually. Statistically, luck runs out. Definite plans, inevitably, make the universe laugh.

***

"Did you bring your clubs?" joked Ted's flight companion, Len Homenuk, while flipping through a British aviation magazine and sipping a cold beer.

"Nope, but that won't stop me from working on my mental game," Ted said, tapping a finger to his head. Then he leaned in toward his associate, adjusting his glasses. "What are you reading?"

"Fly Past. Did you know that the first jet engine to run was a hydrogen-fueled Heineken He S2?" Len rattled off, raising his beer glass a few inches off his tray table.

Ted laughed. "I did not know that."

Then the two men fell into comfortable silence, reading, resting, reflecting, as people do during long rush-hour commutes on the subway—or forty thousand feet up in the air, floating in faith on probable outcomes. Over the past few years Ted and Len had worked together on the feasibility study of the Kyrgyz gold deposit. Finally that June, Cameco Corporation and the government of Kyrgyzstan had signed a joint-venture agreement to develop mine operations in the new republic.

It was July of 1995. Kyrgyz gold was a shiny one-way ticket that many expected to launch the former Communist annex into a glorious, new era of free-market capitalism. That was the official plan, anyway. Discovered back in 1978 during a Russian geophysical expedition, Kumtor was among the top ten biggest gold deposits in the world. But at the time, the high cost and extreme challenges of extracting minerals from a remote and mountainous location had stymied even local interest.

Following independence, in 1991, the Kyrgyz government strategically opened the market to outside bidders. Vigorous interest from the West had kicked plans into high gear, inducing a wave of foreign investment and financial aid into the fledgling republic. The government coffers overflowed. By 1993, several scandals had dotted international headlines involving the misappropriation of national treasury funds by high government officials.

Despite the bad press, Cameco Corp. still wanted in. The Canadian company already held rank as the world's largest publicly traded uranium company, and the Kyrgyzstan project would put Cameco on the map as a major producer of gold. As president of Kumtor Operating Company, a subsidiary branch of Cameco Corp., Len was charged with managing the whole Central Asian mining operation. Cameco had awarded its Saskatchewan neighbor, Kilborn Western, Inc., with a major contract to construct a fully operational mine by 1997. After a series of reassignments, Ted stepped up as Kilborn's new project director on Kumtor, with more than twenty years under his high-sheen cow buckle, working on high-profile projects in the minerals sector.

Ted and Len were aces in their fields. And Kumtor was no small-scale sandbox excavation but a geotechnical feat that would cost $460 million to get up and running—a drop in the bucket, considering that over its lifespan Kumtor would produce more than $8 billion in gold revenues and gild government coffers many times over. By 1997, Kumtor was slated to begin production of 1,500 ounces of gold a day. The multibillion-dollar project would create 120 high-paid positions for Canadian executives and skilled tradespeople in addition to thousands of new work posts for Central Asian laborers.

As the sun set, above the clouds and several time zones away from his home, Ted stared out the window, lost in thought. He twisted the Iron Ring on his pinky finger. Mona would soon be teeing-off with her Tuesday morning ladies, and the kids would be heading off to work and school. Anita was twenty-four now, and she was working for Kilborn in Saskatoon as a document controller. His son, Paul, now twenty-two, was doing his practicum in forestry that summer. Ted felt an exciting rush of adrenaline and involuntarily gripped his armrest; he, too, was embarking on new adventures. Next month, Mona would be joining him in Bishkek and who knew what stories they'd have to share when they next set foot on Canadian soil.

Then, without knowing why, Ted suddenly found himself thinking of all the miners, geologists, and engineers, their spouses and offspring, who made up their large Kumtor clan; many of whom he had yet to meet. The expansive web of this transient and mixed group now extended across borders, countries, and even continents. After three years, Kumtor was no longer simply an ambitious business plan or elaborate CAD blueprint but a living, breathing entity and, like any project, subject to change, chance, and human fallibility.

As the sun set through Ted's window, he reflected on what he'd learned about projects over the years; they weren't simply management equations involving time lines, scope, budget, and quality measures. The human factor was, by far, their greatest resource—that is, if the right team was assembled and working for you. Ted thrived and—like anyone—could falter in this arena: developing people and teams, responding to dynamic human variables in any number of predictable and unforeseen situations. Sentient math. Human calculus. Life's little contingencies.

He had a good team, the best, in fact. But would it be enough to make Kumtor worth its weight in gold? Then without a further thought—about alchemy, airplanes, golf swings, or otherwise—Ted drifted off to sleep. Gold would change Kyrgyzstan, and Kyrgyzstan would change Ted. Had he known what lay ahead, however, he would not have caught a wink on that flight.

LESSONS LEARNED
The more you plan, the luckier you get.

CHAPTER NINE

Master Checklist

The day of October 4, 1995, began at the Kumtor Gold Mine, like any other in recent months. The site was humming with the sounds of motor graders, jackhammers, and earthmovers, echoing their dusty industrial serenade as they beeped in reverse from the open-air pit. It was sheer music to a project manager's ears.

But Ted wasn't there. That day, he was traveling with Mona en route back to Bishkek, after a few weeks at home in Canada. He'd taken the overseas job on the condition that his wife could eventually join him in Bishkek. In Ted's absence, Nick Mills had been running operations at the mine site. He and Ted had met back in Saskatchewan in 1988 on a small mine job for Kilborn Engineering. They'd hit it off and worked so well together they'd been teaming up on projects ever since. Whenever Ted needed a construction manager, if Nick was available, he was Ted's guy.

That day, as Nick went about his early morning site rounds, he took a deep breath and inhaled a pungent mix of cement and limestone. A slight spring picked up his step, and he started to whistle. With a thermos of coffee in one hand and a giant roll of blueprints under his arm, Nick was in his element: alert, on the go, and in the thick of the action. From behind a desk, mistakes got made. But in steel-toed work boots and a tieless collar, this one-man mobile office could see to it that things got done right, on time and on budget; if projects were moving targets—as nothing ever went quite as planned on paper—Nick was always aiming for a running bull's-eye.

An early riser, by 7:00 a.m. he'd already been up for three hours. Juicing up on caffeine, he'd spent a couple of quiet hours going through jobs, tweaking the schedule and prioritizing issues that would require his attention that morning. As manager of a crew of thirty supervisors, contractors, camp administrators, and warehouse leads on-site, it was Nick's job to make sure everyone had what they needed—the right information and materials, and any issues and problems sorted out in a timely fashion—so as not to impede productivity and work flow. With close to eight hundred workers on-site—70 percent of whom were Kyrgyz and the rest international workers from Canada, Turkey, England, France, and the United States—Nick's due diligence enabled site managers to keep the wheels turning while he attended to any emergent matters, either at Kumtor or with the engineering groups in Bishkek and Toronto. Like a well-oiled, high-performance machine, a project's success required ongoing service and maintenance. Nick got a rush out of making things happen.

As the sun rose through the cool mist of an early mountain winter, the construction manager turned up his collar. He was headed to base camp, where a crew was already busy at work putting the finishing touches on some roofing. Whistling as he approached, Nick raised his mug at three crew members.

Henry was a young Vancouver chap they'd brought in from Travco. Floyd and Norm came highly recommended from Alto Construction in Saskatchewan. All three were top-notch workers and perennial team players.

"Hey, there, guys, how she be?"

"Hey 'a there, Nick! If it ain't the cleanest shirt on-site," Floyd grinned, from under the rim of his hard hat. By then, they'd gelled. Everyone at camp got a kick out of their ribs and camaraderie.

"Looks like a storm's coming, boys." Nick glanced toward the east.

"Sure is." Floyd wiped some sweat from his forehead and adjusted his hat.

"Seems O'Henri, here, ordered up some of his crap west-coast weather," said Norm.

"Express post, just for you, Mr. Parenteau." Henri's eyes laughed.

"Should blow in around noon," Nick added, matter-of-factly, checking his watch. "Keep an eye on it. If she gets slick, watch your footholds. We'll make a call if we have to. What's the ETA on window installs?"

"Gonna be wrappin' this roof up in a couple hours," Norm confirmed. "Then we're good to go on them windows."

"Good. I'll get 'em sent out from storage right away." Nick reached for the walkie-talkie hooked to his belt. "Keep it up, guys. She's looking solid. Oh, and I heard from canteen this morning we've got fish 'n' chips for lunch."

Floyd rubbed his belly. "What time is it, there, boss?"

"About that time, Floyd." Nick winked, adjusted his hard hat, and turned on his heels.

"Have a good one," said Norm, raising hand to forehead, in mock salute.

"Cheers." Henri nodded and Floyd took a long gulp from his Tim Horton's travel mug. As the joke went, his loyal companion "Timmy" had crossed four continents with him. The man and mug were inseparable.

"Back at 'er." Floyd reached for a hammer and nails.

"In a while! And watch each other's backs. If she blows in hard, she could get messy."

The guys erupted in laughter. Nick shook his head, already with radio to one ear, and swept the air with his hand in mock disgust at the guys' undying delight in sexual innuendo. Anything out of one's mouth was fair game for their twisted amusement.

Nick chuckled as he headed north toward the processing facilities, less than fifty meters away, where another crew was pounding double-headed nails into stakes with sledgehammers to make concrete slab forms. They wanted to review the engineering specs for building flexible foundations on top of unstable permafrost to clarify a few questions. Nick quickened his step and resumed his whistling.

Phase one site prep was well underway. They were already ahead of schedule, and work was progressing steadily on construction of ninety-five kilometers of main access roads and bridges in and around the mine site. Nick consulted Ted's master checklist. Right from the get-go, his boss had asked for a summary of project workflow distilled into one eleven-by-fourteen-inch page, so at a glance, you could grasp the project and keep check on key components. The guy was an efficiency guru. And Nick had to admit, the system worked like a charm. They were right on track. With the mine's base infrastructure in place, they'd be primed for engineering and design in just a few months. They had their work cut out for them, but with a solid plan, top-notch team and leadership—this job was shaping up to be golden.

Satisfied, Nick checked his watch again. In a few hours, he'd be hitting the road to meet Gary Morris to do a face-to-face hand-over. After three weeks on-site, Nick was going on turnaround, heading back to Canada to spend some time with his family in New Glasgow, Nova Scotia. Since Nick preferred to avoid helicopter flight, he'd given up his spot on the aircraft to another crew member wanting to spend the weekend in Bishkek, and he and Gary had made arrangements to meet at the entrance to the Kumtor site, an hour and a half's drive away, by the town of Tamga on Lake Issyk Kul, just before the road veered up into the mountains.

Nick glanced toward the dark clouds slowly moving in like clockwork from the east, confirming earlier weather reports. From under his coat, he shuddered suddenly. Given the weather, he was thankful that his imminent travels were to be over solid ground. A storm was brewing.

> **LESSONS LEARNED**
> 1. Get your boots on the ground.
> 2. Develop an overall master program that summarizes all project activities in an easily readable format.

CHAPTER TEN

Not Just a Day

By 2:00 p.m. local time, the Kumtor mine site looked like a mud-wrestling ring. Cold rain pummeled the crew and slickened heavy machinery.

Over the past hour, through wind and intermittent wet, Nick had made his rounds checking in on various crews. There were still a couple of hours of daylight left, but with deteriorating conditions working at such high altitudes, upon a shield of permafrost, productivity was dropping and safety and morale were on the line. The decision was a no-brainer.

Nick put the walkie-talkie to his mouth. "Time to pack 'er in, folks."

At that point, he and his administrative manager, Ivo Cervicek, were sitting in a portable office plowing through a couple of construction issues pertaining to various subcontractors. After a few minutes, Ivo took a call, and Nick ducked out, zipping up his rain slick, to oversee shut down.

By 2:45 p.m. the work site was empty and secure. Crew members had returned to base camp or else were waiting for transport by chartered helicopter to the fast-growing capital of Bishkek, 350 kilometers away. Several men were headed home after their three-month stints at camp, while others were taking a few days' leave to spend in the relative civilization of Bishkek. In just a few hours, a dozen of their guys were scheduled to catch the same plane home that Ted had just flown in on earlier that day. Nick would likely cross paths with them at the

Bishkek airport, but he'd be arriving in his own time by car; judging by the weather, which had now stabilized, he was in for a fairly decent road trip.

It was 3:00 p.m. Some slanted rays of sunshine were even poking through breaks in the cumulus clouds.

"Air transport, how we doin'?" Nick spoke in an upbeat voice into the receiver.

"All air transport proceeding as planned. Over."

"Good."

Nick shoved the walkie-talkie back onto his belt. Starting to overheat, he pulled back the hood of his rain slick. "Jeez, Louise. It's heating up!" He was talking to no one in particular, and then consulted his clipboard one more time for good measure. All their ducks were in a row. It had been a great day despite the inclemency; now, to send the boys home, grab his bags, and hit the road.

He was looking forward to the solitude and time to regroup, after a busy run at Kumtor; he didn't need the added spike in his blood pressure from flying through the mountains in a helicopter. Some guys got a kick out of the mile-high experience and spectacular views. Not him. Feeling his feet on solid ground then, Nick was glad he'd given up his spot and that four wheels and a dirt road were in his near future.

At that exact moment, back in Bishkek, Ted was sitting at his desk sifting through updates to the project schedule, which Nick had faxed in earlier that morning. After dropping Mona off at their rental apartment, Ted had taken a quick shower, changed into some fresh slacks, and headed into the office for a few hours. After his time off, Ted felt refreshed and was soon pleasantly reimmersed in work.

Ted wondered how they'd managed on-site. Surely the torrential downpour had hit the camp. So far, there were no emergencies. His construction manager, Nick, was a sharp guy and entirely reliable. He would have sent news down the pipe if anything required his attention. Ted had no qualms about leaving things in Nick's competent hands when he was away. Soon, as per his Day-Timer, Gary Morris would be online, covering for Nick, who was by now probably good and ready for some time off of his own.

Then suddenly, for no reason that he could pinpoint, Ted glanced at the wall clock, whose second hand seemed to be ticking louder than usual, and he felt an itch that he couldn't scratch.

**

In the next quarter hour, three Soviet Mil Mi-8 helicopters carrying human cargo got the go-ahead from air-traffic control and lifted off from the Kumtor Gold Mine site, headed northwest into the winding bowels of the Tien Shan Mountains. It was a relatively short but serpentine trek, less than an hour, door to door, through a jagged maze of alpine peaks and valleys. From four thousand feet on a clear day, passengers could see dirt roads meandering through small village settlements and flocks of Argali sheep grazing in rich green pastures. It was picturesque, but in the wrong conditions, potentially treacherous. Nonetheless, the trek from Kumtor to Bishkek had safely been made dozens of times before. The local Kyrgyz Air pilots knew the rugged terrain and air space like the backs of their hands. If push ever came to shove and there was a system malfunction, it was said in these parts that local pilots could fly the route blind with nothing but a stopwatch and a working altimeter.

Nick felt his gut tighten at the thought. He suspected this was just talk not actual fact. Objectively, he knew that Kumtor's insurance policy regarding flight travel in and out of camp was based on the statistical averages of aeronautic safety; still he'd rather drive. That day, Nick saw the first chopper depart, waved good-bye to the others, and then made haste to prepare for his own departure. He had a five-hour journey ahead of him that afternoon. One he'd never forget. He'd better get on it, he thought.

***

One of the three Kyrgyz Air transporters left Kumtor at 3:15 p.m. It was a textbook takeoff, from start to finish, according to Ivo Cervicek, the administrative manager, who was serving as ground crew loader that day. He watched the chopper cross the Barskoon Pass at 3,800 meters above sea level. Then he meticulously updated his logbook, ate half a Mars bar, and killed some time yakking with a couple of the other ground crew, waiting for word on the third chopper. Soon he'd get word from the radio tower that she was through and clear. Then they could call it a day.

"Any minute now," he thought. Ivo was anxious to pull off his wet gear and hit the camp gym. It was 3:25 p.m.

Two minutes passed. No word. Any second and he'd receive the transmission. Another minute slipped by. Nothing but silence through the airwaves. He checked the power, volume, and frequency settings on his CB. All systems were functioning fine. He suddenly felt uneasy.

"Radio tower, are you there?"

"Yeah, we're here, Ground."

"Any word from EX 25179?"

"Nothing. Something's not right. It's been too long."

"You're right. Shit."

"Stand by, Ivo. Radio tower to EX 25179. Come in, 25179."

Nothing but static. Something was terribly wrong.

By 3:30 p.m., the pilot aboard the Kyrgyz Air Mi-8 EX 25179 had still failed to make any contact with the local radio tower. This standard in-flight safety and communications procedure had never been breached at Kumtor. Ivo forgot his hunger and the itch of his wet gear. For a moment, he stood there in shock, staring up at the sky, where less than a quarter hour ago the last chopper had disappeared into the mountains. This was beyond bad.

Springing into action, Ivo grabbed his CB, radioed the entire site crew, and then bolted back to his office to call Bishkek. He spoke with Rod St. Jacques, the manager of project control, who immediately went to Ted's office.

"Ted! We are missing a helicopter!"

"Did you say—?"

"Yes!"

"What the—?"

When Rod gave him the rundown, Ted's workflow chart dropped to his feet.

"Jesus Christ!"

One second, his head was reeling; the next, adrenaline spiked in his veins. Collecting his thoughts, Ted took control of the situation.

"Do you have a location on the other choppers?"

"Contacting the other pilots now." Rod spoke into the CB he was holding. "Copied. Making contact now. EX 25451, EX 25567, come in...EX 25567."

Ted grabbed a receiver and jumped online with instructions. "Find out their locations. Let's get someone back in there right away to look for the missing craft." Turning to Rod, Ted said, "I've got to call Len."

His associate nodded, nervously.

"Copied," came Ivo's voice over the CB.

"OK, then. Good luck."

Ted immediately dialed his associate. As he waited for him to answer, he took a deep breath. Could this really be happening?

"Hi, Ted. Welcome back!"

"Len, we've got a serious problem. The third chopper heading to Bishkek hasn't made contact. Len, she's missing."

In the next few seconds, air traffic contacted the second charter, still en route to Bishkek, and it immediately turned back. The search began. Word spread like wildfire through base camp that one of the choppers was MIA, and the entire crew converged by the helipad, anxiously waiting for news. There were a number of possible explanations: the chopper could have lost communications, or experienced another system malfunction, and had to make an emergency landing inside the mountains. No one wanted to think of the worst-case scenario.

Time was of the essence. Daylight was limited; temperatures inside the mountains were at least 10 degrees cooler, and oxygen was thinner. No one was certain what survival gear they had on board. At that point, there were many questions but no answers. What the heck had gone wrong anyway, and were the passengers aboard in peril, or even alive? They had to get in there and find them. Any minute now, they could spot them and send in a rescue team. As minutes passed, dread set in. What had befallen their crewmates?

***

Back in Bishkek, Ted and Len were discussing the situation in Kumtor's office. Len had gathered the key Kumtor management team together and was organizing the search and communications both to the Kumtor team and externally. Ted stated that it was necessary to make a call to North America to advise the families of the missing people. Len was reluctant.

"We don't know what, for sure, if anything has happened yet."

"Something's happened. And this is an international incident. We've got to call it in. We can't let them hear it on the radio."

"OK, OK. You've got a point." Len relented. "I'll be there in five."

Kyrgyzstan National employed all three pilots and owned the Mi-8s. Their ground controllers had already received word that one of their craft was MIA and had immediately contacted local police authorities, who'd mobilized a search party. They were now headed toward the mountains in partly overcast skies.

Ted hung up and stood by anxiously waiting for an update. Finally, the phone rang. At the same time, Len flew through the door out of breath.

"Any word?"

They were monitoring the search, and through the line Ted was able to hear and connect with the Kyrgyz pilot, Amar, who was flying the returning chopper and now reentering the Tien Shan Mountains from the other side. Ted hit speakerphone, and he and Len joined the conversation.

"How is visibility, Amar?" Ted asked calmly, his forehead deeply furrowed.

"Outside was fairly clear. No problem. But the clouds inside have dropped and formed a blanket inside the mountains. It wasn't like this fifteen minutes ago. We have only one mile now of visibility at a thousand feet above clouds and five hundred below. I can't see much ground now. Stand by please."

"OK, Amar. We are standing by. Take 'er easy." Len's voice was steady.

He pressed his fingers into the desk as they waited, and the two managers silently exchanged a tense look. As president of Kumtor Operating Company, Len was in charge. Together, he and Ted, would handle this. But suddenly, Amar's voice sent terror ripping through them.

"Chyort voz'mi! Holy God!"

"Amar!? Are you OK? Do you see the chopper?" Ted asked, pulse racing.

"No. No. I didn't see. But there is some lightning in the vicinity. We must stay back and wait until this system clears a bit."

Ted suddenly wondered what were the chances of being struck by lightning but then pushed the thought from his head.

"OK, standing by. We are right here with you, Amar." Len stared intensely at the phone, as if with enough focus he could will himself across 350 kilometers of mountain terrain over to the mine site. Ted assiduously took notes in his Day-Timer, in case they needed it later.

The sky was growing darker by the minute. The Kyrgyz search-and-rescue team had moved in and was hovering around the mountain pass perimeter looking for clear pockets to maneuver through.

Then through the airwaves, Amar's boss at Kyrgyz Air ordered him to turn back toward Bishkek. The police would take over the search from there. Amar protested. The missing pilot, Aybek, was not only Amar's colleague but also his longtime friend. They had flown together through these parts for more than twenty years; their families frequently ate Sunday dinners together. He couldn't leave his friend out there. His protest was in vain; the boss's order was absolute, and Amar's fuel gauge was getting low. He knew they couldn't risk losing him also. And so Amar dutifully turned back to Bishkek.

At that moment, Amar said a silent prayer for the missing passengers and crew. Ted and Len could hear his chopper veer around then fade into the distance. Under his breath, Amar repeated the words, "Kuday, kuday. O, kuday, kuday." On the other end of the line, Ted and Len were silently praying in their own way for a safe outcome for their missing team members. It was 5:39 p.m. local time.

In a few minutes, Nick arrived in Bishkek, oblivious to the horrible turn of events, and decided to stop and grab a drink at the local bar located on the main floor of the Euchkun building, where Kilborn and Kumtor's project offices were located. His plane to London was leaving at midnight, and he had a few hours to kick back and relax, before heading to the airport. As Nick took a sip of the cool, frothy pint placed before him, flipping through a week-old British newspaper that was lying around, he noticed a couple of guys from the office upstairs and wondered why they both wore stunned expressions on their faces. Catching their eyes, one recognized Nick and jumped up, approaching him.

"Hey, Nick." It sounded like a question.

"Hey there, you fellas just off work?"

"Did you hear about the missing chopper?"

"What? No! What's happened?"

His friend joined him, and both filled Nick in: a search was in progress for one of the three helicopters, which had failed to communicate with radio control that afternoon during its routine flight to Bishkek.

For the second time that day, Nick shuddered. He set down his beer, feeling stone cold under his wool sweater, and quickly paid his tab. Then thanking the guys for the update, swiftly made his way to the upstairs offices where he knew he'd find Ted. What the hell had happened?

Nick's name had been on the flight manifest for travel on one of those choppers. Ken Hermann had taken his place.

***

Back at Kumtor, the slanted October sun, still partly obscured by cloud clusters, slid behind the jagged mountains, shrouding everything at camp in black. The temperature plummeted, and team supervisors and crew moved indoors. Nobody spoke or ate much that night. The shock had obliterated their hunger and shaken each team member to the core. They were all too conscious of the fact that it could have been any one of them up there.

Meanwhile, half a world away, a golden sun was just peaking up over Saskatchewan's endless horizon, where several of the missing crew members' family slept soundly, still oblivious to the crisis unfolding.

***

At 6:00 p.m., Ted made the call to Alto Construction, got switchboard on the line, reported the emergency, and was put through to the president's home residence. It was one of the most difficult calls he'd ever had to make. And this was only the beginning. Meanwhile, Len remained online with the radio tower, following the minute-by-minute search as it unfolded. Suddenly Nick flew through the office door. Seeing his colleague, Ted nodded and approached Nick.

"You heard."

"Yeah. Any word?"

"Nothing yet. A search is in progress."

Looking at his wristwatch, Ted ask Nick when he was flying out.

"Ted, about that, I'm more than willing to stay and be whatever help I can be."

At this point, Nick had been away from his family for three weeks. Yet under the circumstances, he wouldn't hesitate to cancel his flight and stay put. Ted placed a hand on his shoulder and spoke calmly.

"Look, Nick, if everything is OK, you would just have wasted your time off. And if the worst has happened and it's crashed, what are you gonna do? Go ahead."

Nick nodded. "OK, but if there's anything at all..."

"I'll let you know. Now go home and be with your family."

Without another word, Nick left and drove straight to the airport. When his plane took off toward London later that night, there was still no word on the missing charter. Little did Nick know that because his name was on the official flight manifest, Human Resources in Saskatchewan would call his wife in New Glasgow, Nova Scotia, along with all of the families of men headed out on rotation, to notify them, quite cryptically, that there was going to be an announcement. For six hours, his wife, mother, and kids would live in fear for his safety wondering, along with the others concerned about their husbands, fathers, and sons, if they would ever see him again.

***

After hours of tense maneuvering and waiting, the search was called off just after dark due to a combination of risk factors. Reduced visibility and unpredictable lightning strikes in the mountains made it unsafe for any aircraft to navigate the cloud-obscured peaks after dark.

"We'll get right back out there at first light," Len said firmly, trying to buoy the others. They were doing what they could from a distance, but nonetheless it felt inadequate, given the circumstances.

"I'll update the rest of the crew," Ivo reassured them. "And I'll call as soon as we hear anything." Then there was nothing left to say, so they hung up.

Ted and Len spent the next few hours working through various protocol scenarios. They knew that local media had already been alerted to the situation. It was only a matter of time before international news outlets caught wind of it. By midnight, their phones were ringing off the hook. Together, they'd written an official statement and faxed it to Cameco's head office in Saskatoon. Len got their external affairs person to handle all press calls. Until they knew what they were dealing with, no one was allowed to talk with anyone on the outside.

Aboard the missing aircraft were twelve passengers, all Cameco contract employees, and three crew members, including the pilot. In total, nine

Canadians, two Brits, a Turk, and three Kyrgyz nationals were missing. No one in charge at Kumtor slept that night.

As Ted and Len saw the sun rise through the window of their third-story Bishkek office, the burden of responsibility weighed down on them like a thousand tons of iron ore. At this point, neither of their wives knew what was going on. It also hadn't occurred to Ted to call his kids at home, who could possibly wake up to news reports of a missing helicopter and crew at the Kumtor Mine in Kyrgyzstan.

As they waited for the search to resume, Ted turned to making notes in his Day-Timer, documenting the most minute details of the last twenty-four hours—a precise time line of events, decisions, dilemmas, action steps, and unknown variables, anything to keep occupied and help put some order to the chaos and uncertainty of the unfolding situation. No one wanted to think of the worst but it was becoming increasingly difficult to avoid it.

***

The search resumed at first light. By 6:00 a.m. local time, police helicopters were circling the clear azure skies like determined birds of prey, searching the rocky gray sierra. Once again, minute by minute, Ted and Len were following the search, on the phone and via some old CB equipment they'd rigged up in the office. Meanwhile, on the ground, all mine work was suspended. The entire Kumtor team was on edge, waiting for news of their missing crewmates. By this point, everyone presumed the worst, but no one was willing to cut hope from a dangling thread and say it.

It was only a matter of time before the dreaded message came through the airwaves. The pilot shrieked when he saw the wreckage.

"Oh Kuday! There it is! I see it."

Len jumped up from his seat, and air traffic copied the transmission to confirm its accuracy. Then Len spoke slowly and deliberately to the search pilot: "What exactly do you see?"

"It has crashed in the valley. Oh Kuday, I can see bodies from here. We are going in for a closer look. There could be survivors who need our help. Stand by."

"10-4. Dammit." Len's head dropped as the reality set in.

Ted swallowed hard and sat bolt upright and motionless, as he waited for further word from the pilot, anything to help understand or explain the horror of events. As they anxiously waited, search-and-rescue craft circled within fifty meters of the crash site, hovering over the smashed remains of the Mi-8. Had they been there, Ted and Len would have seen burnt metal and bodies severed and strewn around the exposed belly of the mangled chopper.

"Oh Kuday Kuday." the pilot's voice broke again.

Dread and disbelief filled the silence.

"Oh, God," Ted's whispered, under his breath. "This isn't happening." He cupped the base of his skull with one hand, as if to fortify himself against the deluge of unwanted news. His throat tightened, and Ted had to steady his voice as he spoke.

"Are there...any signs of life?"

There was a pause. Then came the pilot's prostrate reply. "No. None. At least not from up here."

No one wanted to give up the last filament of hope.

"We've got to land and get in there."

"Do it." Len spoke urgently.

A few strained minutes passed and more waiting. Ted and Len didn't speak. There weren't words for the depth of devastation they were trying to process. Finally, the pilot's definitive words cut through the radio static, confirming everyone's greatest fears.

"They are all dead."

The next few seconds were a blur. Then through a fog of swirling possibilities that slammed back into one harsh, cruel reality, Ted heard the pilot's voice from very far away saying that his fuel gauge was low and that within minutes he would be approaching for landing at Kumtor's helipad. Ted and Len looked at each other and, without further delay, sprang into action. This was on them. And they needed answers. Len bolted toward the door.

"I'm going over. A chopper should be ready."

They'd tried to imagine and provide for all possible contingencies, and so they'd arranged for transport during the night. As they'd discussed, Len would head to Kumtor to implement crisis management procedures. They'd need to mobilize a recovery party, file police reports, and assist with the investigation; furthermore, the crew and management team would need support and counseling in the coming hours and days.

Ted would remain in Bishkek and handle all communications; brief Cameco and Kilborn head offices; contact all their lawyers; and launch a private inquiry so they could find out exactly what went wrong out there. Then there would be telephone calls to the families and public statements to the press, and they'd need to bring the recovered bodies home.

"If they've crashed, there's gonna be a national investigation. We will need to conduct our own independent inquiry," Len had said, in the wee hours of the morning. Before leaving, the next morning, to catch a ride to Kumtor, he reiterated their plan.

"We need to get some answers, Ted, as to what the hell went wrong up there. Those were our guys. Until we know what went down, this is on us."

This is on us. The impact of those words was like a dull, hard blow that echoed deep inside Ted's cerebellum. There would be lots of questions. And they had the families to answer to.

Ted spoke up. "We'll get answers, Len. It's on us. We'll get answers."

Without another word, Len left and Ted sat staring at the door. He drew in a heavy breath. Dread and then extreme exhaustion washed through him. Ted thought of Mona and realized she'd be wondering why he hadn't come home and reached for the phone to call her.

How do you tell someone else's wife, fiancé, girlfriend, family that their loved one is dead? In Ted's nearly twenty-five years managing projects, he'd never had a single fatality on his watch. It was a point of pride, and he'd often thought, matter-of-factly—usually whenever he had to wade through one of those thick procedures manuals or attend yet another emergency protocols management seminar—that human resources' whole mining safety program was a tad overdone. Now he realized all the hoopla was for moments like this: when the illusion of total security suddenly vanished into thin air, when he had fifteen bodies to deal with and no solid answers to hold onto and explain what went wrong.

It was October 5, 1995. On that life-altering day, when he learned his crew members were dead, a part of Ted that had put his faith in probabilities and safe outcomes for years died a swift and sudden death. Then, barely above a whisper, Ted said to himself: "What were the goddamn chances?"

Mona suddenly picked up. His eyes brimmed and his stomach twisted in knots when he heard his wife's voice.

"Ted, where are you?"

"There's been an accident. We lost fifteen men."

LESSONS LEARNED
1. Maintain your composure no matter how serious the crisis.
2. Don't make light of serious situations.
3. In crisis situations, it is mandatory that the lines of communication are clear.

CHAPTER ELEVEN

Safety and the Law

"He did what?? Cut off his thumb!? And his finger?! Jesus!" Cell phone to ear, Ted was pacing around Jim and Beryl's backyard. So much for their quiet, relaxed Tuesday morning.

Moments before, the two cousin-in-laws had been sipping coffee, buttering toast, and leisurely flipping through the pages of the *Tuesday Kelowna Courier*, like a couple of "normal" retirees settling nicely into their pre-golden years. Ted was sixty-six and Jim had turned seventy-nine that year. Despite repeat assurances to their patient and forgiving wives, neither had officially retired; nor was either quite ready. Each still got a thrill out of the game and ran his own consulting company, with a more or less arm's-length involvement in a handful of lucrative, high-profile international projects.

Beryl and Mona had long ago accepted that work interruptions were an inevitable part of their lives, even on family holidays and vacations. That morning the ladies had knowingly left their husbands on the back deck to their own devices and happily headed into town to grab lattés at Starbucks, take a walk along Bernard Avenue, and then maybe duck into a new outlet mall in town.

Times had certainly changed in the men's respective industries—strategic business had shifted, management protocols were ever evolving, and from the human resources and insurance standpoints, safety practices and employee retention were more important than ever. But some things remained the same. Accidents happened. Within minutes of their wives' departure, Ted's phone rang.

It was Trevali Mining Corporation's CEO Mark Cruise. Ted sat on their board of directors and was the safety committee chair. He was calling to report that one of their carpenters, stationed in Santander, Peru, had just cut off his finger and thumb.

"He was explicitly instructed not to use the power saw," Mark told Ted. Apparently on that specific day, this fellow was without supervision. He did not have the forbidden tool properly guarded when he tried to prove to himself that he could handle the equipment.

"And now he's minus a finger and thumb," Ted said. "Christ!"

From the deck, Jim looked up from the paper and shook his head knowingly. He'd seen his share of senseless accidents over the years. In the mid-1980s, Jim had been working on an oil project off the coast of Newfoundland. There was a guy on the rig who, in hindsight, had probably exhibited signs of stress and depression for weeks. But mental-health clearances and mandatory crew education around identifying the telltale signs of such potential hazards or dangers had yet to be integrated into workplace safety procedure. In those days, if you were fit in body and had your necessary papers, you were fit to work out there on the rig. One afternoon, plain as day, this guy was squeezing a Holy Bible tightly in his hands when he walked right off the end of the rig into the ocean, never to be seen again. They sent divers in after him, but a body was never recovered. Jim wasn't on board during that work stint, but this fact never stopped him from replaying the horrible tragedy in his head over and over again for years to come. A few of the guys were so shaken up that they required counseling, and at least one asked to be reassigned to a land division.

Ted finally hung up the phone, after discussing with Mark how they should handle the situation. He made his way back up to the porch, looking wired and perturbed. Jim set down the paper and crossed his hands in his lap.

"The guy wasn't cleared to use the power saw, but he did anyway. Stupid, stupid, stupid." Ted shook his head and plunked himself into a patio chair, slouching.

"Now he's ruined his career, and on top of that, he's disabled. That is...really too bad." Jim's words, as usual, were measured; his clear blue eyes calm and attentive.

Ted just nodded. A minute passed in which he scanned the length of the deck floor, mulling things over. Situations like these were awful, plain and simple; they also created more work. It was bad enough that the Peru project was already wracked by local union issues, not to mention was behind schedule and slipping off budget. Now there would be Workers' Compensation Board claims to deal with, supervisor reviews, replacement and reassignment issues. Disaster management was a part of project work that Ted didn't miss; fortunately, the main on-site management guy, Paul Keller, would handle most of the logistics. However, Ted would oversee the matter from arm's length, and it didn't make it any easier being on the receiving end.

Ted took a swig of coffee. It was a Tuesday. No day was a good one for such horrible news. But calls like this were just part of the job. You dealt with what came up and then tried to let it go, as hard as that could be. He'd learned that from Kumtor. He glanced at the clock hanging on the yellow stucco wall of the half-enclosed patio. Still too early for happy hour. Maybe a walk around the block, and then the girls would be home and they could crack open a bottle from their Okanagan Reserve collection. The wines from these parts were top notch; he and Mona had picked up a few bottles at Quail's Gate vineyard on their way into town.

Suddenly, Ted's eyes brightened, as a thought about John Somers popped into his head. Back in the late '80s, at the end of a long workweek, his former boss and mentor would open up the bar in his office, and they'd all have a few drinks. John called it "No Decisions Fridays" and enforced the policy with spirit, but there was also a serious side to this practice of kicking back and detaching from the workweek. It was a preventive safety measure. A few years back, the Vancouver Kilborn engineering exec, Bob Hosey, had been shot dead after firing one of his employees on a Friday afternoon. They'd all taken a lesson from that horrible incident and never made an important decision regarding employees on a Friday. John knew how to take a negative and turn it into a positive.

A wrinkle grew across Ted's forehead as he refocused on the issue. What could he do now, after the fact, once the damage was done? The problem with policy, Ted thought, was that the problems precede the policy, and even when policy was in place, disaster could still strike. Over the years, the mining industry had been shrink-wrapped in many layers of policy: safety protocols, mandatory supervision, employee clearances, and best practices were all in place to help to prevent situations like these. But all it took was one lapse of judgment or plain bad luck, and those protocols were rendered useless, and someone's life and livelihood were on the line.

Ted's back straightened. Days like this made him think of October 5—and how much he hated that day. After more than a decade, the Kumtor crash and its long drawn-out aftermath were never far from Ted's thoughts.

Safety and the Law

In terms of performance, the project had been deemed a major success—budget, scope, schedule, and productivities—and after Kumtor, Ted's career had really taken off. When people were looking for a director on a large-scale project, they came to Ted. Kumtor launched his career into a whole new stratosphere: bigger and more international projects, fatter pay checks, more opportunities, and greater capacities to influence the direction and quality of major, multibillion-dollar mining projects. Kumtor was one of the great thrills and challenges of his career, and he was proud of his accomplishments. On Kumtor, Ted learned how to remain calm under pressure. He learned how to handle the stress, including the crash aftermath and subsequent ballooning costs, mad clients and disgruntled staff, without getting lost in the details or spitting the dummy—an Australian idiom for having a tantrum, meaning, literally, a baby who, having spat out his dummy (pacifier), is screaming.

But the golden success was tarnished by the human losses, and Ted never quite got over it. That first fatality was a game changer. It rocked his world and his status quo. Before Kumtor, he's often thought, "What is all this crap about safety?" Up until that point, he'd been lucky to work on projects that were free of any life-threatening accidents. Kumtor had popped the bubble of his illusion of safety.

And so, after the accident, safety became Ted's passion and preoccupation. He began to scrutinize every job site he stepped onto and would carefully screen new hires to make sure they had their wits about them; over time he developed a sixth sense for whether a site was secure or not. Within a couple minutes of standing on a site, Ted could feel it in his gut and bones, if something wasn't right. Word got around; if you wanted a secure site, the best possible team, and chances of project success, Ted Bassett was your man.

After Kumtor, Ted's priorities not only shifted to crew and site safety, but his very definition of project success gained a depth and maturity that only time and tragedy bring about. He was humbled and wizened. Never again did he take safety for granted. He now believed that scope, schedules, budgets and the right assembled team were moot if you couldn't keep everyone safe.

Before Kumtor, it was all about the project. Afterward, for Ted, people came first. Period.

After Kumtor, in his personal life, Ted came to value his relationship with his family and his wife even more, as of the initial five Canadian personnel on the ground in Kyrgyzstan who went overseas on the project, three ended up getting divorced shortly after. After Kumtor, Mona had joined him overseas, and they'd stuck together.

Seeing the cloud hanging over his friend's head, Jim broke the silence. "You know, there will always be guys like this fella who, no matter what you do, just won't listen. You do what you can, to secure the site and put protocols in place, but these things just happen."

"It's true." Ted looked at Jim, an old, dull pain in his eyes. "You never quite get over the first ones, do you?"

"No, sir," Jim sighed. "You sure don't."

Ted rubbed his temples with the thumb and index finger of one broad hand. "But you never quite get used to dealing with any accident thereafter either."

Jim half shrugged and remained silent.

Kumtor had dragged on in the courts for a decade. Answers came slowly, bringing a measure of peace and closure to those touched by the tragedy. But the answers were followed by more questions, some of which would be forever unanswerable.

Less than a week after the helicopter crash, fifteen bodies were recovered from the wreckage and the remains were sent home. On October 15, 1995, John Somers and Nick stood among mourners at a memorial service in Saskatoon, held for several of the lost crew members. Ted remained at the work site in Kyrgyzstan to manage the personnel and issues there. For those most closely touched by the crash, there was only life before Kumtor, and then life—and all of its holes—thereafter.

In the early days following the crash, the Kyrgyz police had offered bad weather and poor visibility as a possible cause for the fatal incident. It took six months for the official investigation results to come in. The final verdict: human error was to blame for the crash that had claimed the lives of twelve Kumtor employees and three Kyrgyz crew members during a return charter to Bishkek from the mine site, located in eastern Kyrgyzstan. According to the report, that fateful afternoon the crew had strayed from the approved flight path and should have turned back to Kumtor when they encountered lightning and poor visibility along the route.

It was a cloudy day and the chopper was only good flying at a maximum of 4,500 meters. The crew knew they were not supposed to fly in the clouds. Once inside the Tien Shan Mountains, the craft had almost cleared the pass but just before turning north toward the lake, reportedly had hit the last mountain, never to emerge out of the other side. The tentative conclusion of the investigating team was that the crew could not see and therefore resorted to turning on a stopwatch; in other words, they were flying blind through clouds and in their calculations may have forgotten to factor in a tail wind, perhaps flew too high or misjudged the timing for the turn. No one would ever know for certain.

Blame was, of course, laid in many directions. As incomprehensible loss often goes, once hope vanishes and prayers are said, disillusionment and anger set in. The deceased's estates sued Cameco and Kilborn for $20.7 million, and of Cameco's three insurance policies covering flights in and out of the mine site, two denied coverage, and so Cameco, in turn, sued their insurance companies. More than a year after the crash, Ted found himself in a discovery room, supplied by the court of Saskatchewan, seated across from the widows of his dead crew members. He would never forget the experience. The air was thick, and minutes passed like hours.

Everyone had a copy of Ted's meticulous Day-Timer notes open in front of them, and the claimants' lawyers were questioning him.

Ted had never been more uncomfortable in his life. But as he sat there, under scrutiny, he reminded himself that his malaise was nothing compared to the families' losses. They were entitled to this hearing, and so he took a deep breath and put his own distress into perspective. Ted had memorized each stark detail that he'd recorded of the minutes, hours, and days surrounding the fateful incident. As they sat there, hour after hour, he let his eyes glide over and over the words in his notebook. This helped him avoid making eye contact with the women sitting across from him, whose faces told stories of inconceivable suffering.

Words he could handle. They were all pieces of information, facts and points of view that they could use to help shed light on the situation and defend their position. In the spaces between the words, however, the human losses haunted Ted, especially on the afternoon that one of the claimants showed up for the hearing with her grown-up daughter and grandchild in tow.

***

Ted shifted uncomfortably in his deck chair as he recalled the moment he made eye contact with that child's mother. Her face was forever burned into his memory.

"I told my lawyer, Sean, that I didn't know if I could do it." Ted spoke quietly.

Jim just listened, hunched forward in his deck chair. Without Jim saying it, he knew they were talking about Kumtor. He had all day for his friend.

"You deal with what life throws at you," Ted continued. "But sometimes you'd rather not deal. That accident was on my watch."

"What did your lawyer, Sean, tell you?" Jim asked.

Ted paused looking up, as if reaching into the past for his lawyer's exact words. "I'm there, looking after the interests of my company. The worst part was having the wives of the guys who died there. But I couldn't talk to them. I couldn't really say anything to them. Only immediate family members were allowed to be present, but then that one day, after a small break in the proceedings, the daughter of one of the wives showed up with her six-year-old. That was one grandchild who would never have the joy of knowing her grandfather—and there were others."

Ted paused, and Jim waited for his friend to continue. "I turned to Sean and told him I didn't think I could handle it. And that's when he said to me: 'The daughter is not specifically named in the court documents, so technically she cannot attend the discovery.'" At Sean's request, she was removed from the room.

"You did what you thought best to get through it," Jim said, nodding thoughtfully. "That's all you can do."

Over the years, Jim had learned from his share of on-the-job challenges.

"You've gotta take the lessons and move forward. That's what my first manager taught me. Back in the early '60s, when I initially started working in the oil patch, we had a number of projects on the go, and one of them was a sour-gas gathering system of pipelines in southern Alberta." Jim studied his left hand and then flipped it over, rubbing a callus, as he continued his story.

"This was before offshore, you see. I was still wet behind the ears, just a few years out of school. We were pushing the technical envelope but we had a top-drawer guy advising us." Jim unconsciously licked his finger and ran it over his eyebrow.

"At the time, we were still learning how to use the right kinds of materials in those situations, with highly corrosive gases. You couldn't use ordinary steel. It was too brittle."

Ted leaned back in his chair, listening to Jim's story. "Was this a sulfur plant?"

"That's right. It was one of our first sour-gas processing plants. It was a big boom year for us. One day, one of the gaskets on the pipeline leaked, and the gas impinged on a bolt, which was not made of sour-gas-resistant material, and we had a huge leak. We had to evacuate the surrounding area."

"Oh, yeah. I remember that."

"They shut down the field. It was contained, fortunately, and they brought all the locals back. Well, our production manager, he wanted to see all of the engineers. We paraded into his office. He only said one thing to us, and I will never forget these words. He said, 'If the shoe fits, wear it.'"

"If the shoe fits, wear it," Ted repeated.

"That's right. Even though we were at the cutting edge of using these materials, there was no allowance for making a mistake like that. We trusted that the gasket wouldn't leak, but there was no second line of defense. He wanted us to understand where the bar was."

Ted chuckled then, recalling what his former boss once said. "That's another thing John Somers used to say to us: 'You might as well aim high, 'cause you're gonna hit low enough anyways.'"

"He's a realist, Somers." Jim nodded. "You've gotta strive for your best though."

"It's serious business. You make a mistake, and people can get hurt, not to mention government regulation starts piling up on you."

"And then you can't move." Jim paused and said, matter-of-factly, "So, going back to my first manager's philosophy, the moral of the story is, if something applies to you, take it to heart."

Then thoughtfully, Jim added, "And know that you can't always come out the other side smelling like a rose."

Ted cocked his head to one side, reflecting on his cousin-in-law's statement. Jim was on a roll.

"Take Turkey, and the Kurdish rebels and those PKK militants who were shooting at us, for instance. It was just one of the hiccups along the road. You just have to keep a level mind and keep on going. You try to set up so hiccups don't happen, but they inevitably do. We got in the middle of the politics of the country. What do you expect?"

Jim was sitting on the edge of his chair, arms thrust in midair.

"You just keep an objective mind, so you can deal with the hiccups as they come up, and play it smart. Due diligence. What else can you do?"

"That's right," Ted nodded. "This thumb guy…" he began.

"This thumb guy, he knew better. But where were the safety managers when this happened?"

"Sitting in their offices apparently."

"There you go!"

"There you go."

"If you've got continuous presence of management, the odds of someone consciously breaking the rules are less."

"They essentially need babysitters."

"Maybe so, but if management is sleeping in the corner…"

"Someone loses a hand."

"Better safe than sorry."

Ted clapped his hands together once. "You know, talking about making stupid mistakes—last week, my eaves trough was overflowing with all the rain we got, so I set up a ladder against the house, and I hooked the hose to the ladder. Well! There was a kink in the hose, and so without thinking, I gave it a pull, and that ladder fell just missing me by a foot. I thought, 'Jeez, I just did something really stupid. I tell the guys not to do that. But I just did.'"

Jim chuckled. "We all have our moments." Then stretched out his legs.

Ted ran his index finger along his jaw line. Kumtor had felt like several lifetimes. In spite of himself, Ted cracked a smile.

"Funny thing, well not funny, really, but you know, in the last fifteen years, since Kumtor, I've seen at least a dozen résumés cross my desk from guys claiming to have been project managers on the Kumtor project."

"You're kidding."

"There were only three, and I knew each one of them. That would make me really mad. I'd see that and think, 'No, I was there. You were not.' Every last one of those applications got tossed in the trash."

"You don't want those ones on your team."

"I think it all boils down to two things: people and training."

"Don't forget the X factor. You never truly know how she's gonna go."

"True enough."

Ted leaned back in his seat and sighed. To some extent, they were all flying it blind, trying to make the best call on situations that could, some days, be as changeable as the weather. Accidents just happened. You could do your best to mitigate trouble, minimize risk; write mountains of policy and train up the yin-yang; but still, accidents happened, because—plain and simple—people happened.

Maybe your guy chooses to turn on a stopwatch or to try his hand on a piece of equipment he's not cleared for; at a certain point, it's out of your hands, entrusted to that indispensable yet hard-to-nail-down-thing called good judgment, sharp wits, and plain, old good luck.

Kumtor had happened. The reality of what had taken place that day, way up in the Tien Shan Mountains came in waves and sank in by degrees; after multiple investigations and court proceedings, stretched out over many years, at long last there was a ruling, compensations were awarded, and the Kumtor docket was finally closed. In 2005, Cameco incurred a cost of $6.4 million to be fully indemnified from claims in connection to the Kyrgyz helicopter crash. Then it was all over, on paper at least. Justice had been served. Hard lessons learned. But for each individual uniquely touched by the crash—wife, girlfriend, parent, sibling, child, future offspring, or crew member—life was never the same. Death's brush left an impression that could not be erased or undone. Each had their own Kumtor.

***

Just then Jim's phone alarm went off. Ted and Jim looked at one another. Then Jim winked, breaking the heavy silence.

"Well, what do you say? A bit of a walk to stretch out these poor, old legs?"

"A walk sounds like a good idea," Ted agreed, jumping up.

On his right leg, Jim wore a knee-high therapeutic sock to help with circulation, due to an accident of his own a few years back. It came close to an amputation and led to the end of his physical work offshore. He needed to move every hour or so to keep the blood flowing. Beryl had him well trained, having preprogrammed little stretch breaks into Jim's phone.

"Squirrelly did that." Jim laughed. "She calls it my preventive medicine. You know, the insurance companies these days won't even give you coverage unless you're doing your exercises and getting your BP checked regularly."

"They're managing their risk."

"That's right. After my heart stent, Beryl and I even had to take classes."

"Classes? For what?"

"For tools to manage your own health—diet, fitness, and so on."

Jim patted his chest. "Personal responsibility. Actually I've never felt better."

"Everyone's moving toward accountability these days."

"The system can't afford it."

Ted considered his employee who was short two digits. "Neither can we."

Jim nodded silently, wheels churning in his head. He clapped his hands together. "OK, I'm going to go put on my runners, and then we can take a tour around the block. I'll meet you back here in five minutes?"

"Sounds good."

Jim stood up with some effort and, gathering their empty mugs, headed into the house, leaving his cousin-in-law on the porch with his own thoughts. For a moment Ted stood there, unmoving. He was thinking of Jim's old boss's advice: "If the shoe, fits wear it." But work wasn't on his mind anymore.

He was remembering one Halloween, a long time ago, when his daughter, Anita, was six or seven years old, a decade or more before Kumtor was ever a word in his lexicon. She was dressed up as Cinderella from her favorite Disney fairy tale, and they were just about to head out the door to go trick or treating. She was sitting on the edge of the stairs, all done up in her pink and white gown, and her brother Paul was putting her shoes on for her, acting like the storybook prince. Mona was standing by with their coats.

"If the shoe fits, then it's yours!" Paul had proclaimed.

Anita giggled and squealed with delight as the slipper slid onto her little foot. "It fits. It fits! It fits! Now quick, before I turn into a pumpkin!"

Ted had recorded the whole thing on the family camcorder. Standing on the porch, Ted smiled sadly. He'd have to dig up that old video for Mona to watch. It was so long ago. Yet some memories never fade.

The shoe had fit, and so he was wearing it.

> **LESSONS LEARNED**
> 1. Sometimes shit happens anyway.
> 2. When human lives are at stake, you can never do enough due diligence.
> 3. You have to agree on procedures up front on any safety and logistical matters.
> 4. In a crisis situation, clear communications and clear responsibilities are vital.
> 5. When you're under severe stress, make sure you take a time-out to clarify your thinking.
> 6. If a piece of the project goes sideways, take responsibility: "If the shoe fits, wear it."

CHAPTER TWELVE

Finding Balance

"Speaking of waiting around for Jim: Did I ever tell you about the time I won the trip to Hawaii?!"

"No!"

"Tell us."

Beryl was yet again teasing her husband for his slowpoke tendencies. Mona and Ted laughed, sipping their wine, as Jim shrugged with a bemused half grimace on his face. She'd told the story at least a dozen times before, but word for word, like the fine oak-barreled wine they were drinking, her little yarn got better with age. Arms bent with enthusiasm, Beryl beamed. Her velveteen cheeks had a pink hue, and she held a glass of Gray Monk Chardonnay in her left hand.

Jim sat back in his seat, crossing his arms, and winked at the others, as his wife of sixty years launched into her delightful story, which was equal parts romance, heroism, and good-natured reprimand. It was happy hour in the Cooke's backyard. The girls had returned from their outing just as their husbands arrived back from their afternoon stroll. The sky was a cloudless blue, which reflected in Jim's clear eyes.

"Well, as usual, Jim was away working on a rig, somewhere—I dunno—in the middle of the ocean," Beryl began. "I hadn't heard from him—in weeks! Not one call." She paused, folding her hands in her lap, and looked around the table.

Jim couldn't resist filling in a few, small details. "This was back in the mid-1980s. We were working for Mobil on a semisubmersible rig, marine and drilling operations out in the Atlantic Ocean. There were one-hundred-plus people on board."

"Well, that doesn't matter to the story. More importantly, I hadn't heard from him in three whole weeks!" Beryl batted her eyes and shook her head in mock exasperation. "I was so depressed."

Ted grinned, leaning on the back legs of his chair. Moments like these, it was hard to believe that more than fifty years had flown by. His cousin hadn't changed a bit in spirit since they were teenagers, he thought affectionately. Beryl regaled them with her funny story.

"One day, when our girls were all in school, I went to the doctor's. I had the worst PMS! I remember bawling to him; he must have thought I was just some crazy housewife. I think he prescribed me some Midol, thank God." Beryl put a hand to her forehead, melodramatically. Mona listened, smiling and sipping her wine.

"On my way home, I was still feeling horrible. I had to stop to pick up some milk at the store. While I was at the counter, I decided to splurge on a lottery scratch ticket to cheer myself up, and I actually won! I'd never won anything before in my life! But that dreaded day, when everything seemed wrong, I won a trip to Hawaii!"

"You called Aunty Vera right away, didn't you?" Jim played along casually.

Beryl's right hand flew to her hips. "Well, who else was I gonna tell? I waited and waited for you to call."

"I did."

"Finally." Beryl shrugged, looking back at Ted and Mona, shaking her head. "When Jim finally called, two days later, I answered, and he said hi, just like he'd talked to me yesterday."

Mona giggled.

Hand held like a phone, Beryl recalled the conversation like it had taken place the previous day, not a quarter century ago. "He said, 'What are you doing, Squirrelly?' 'What am I doing?' I said 'I'm going to Hawaii.' He says, 'Who with?' I say, 'Tom Selleck.'"

"Ha!" Ted slapped his thigh. They all laughed.

Beryl looked directly at Jim, a split-second glance that summed up years of conversation between them. "Well, after that, he knew I'd had it."

Jim shrugged his shoulders and replied matter-of-factly. "So I flew home."

Beryl's face grew indignant. "He didn't just fly home. He got a helicopter to fly to the rig early in the morning in the middle of the ocean during a bad storm to pick him up and bring him home. He could have died!"

Jim grinned sheepishly, winking at his wife. "You gotta do what you gotta do."

"Oh my God, that sounds scary—but a little romantic too," Mona conceded.

Then Jim's face grew thoughtful. There was another part of that story that had stuck with him for years. It had been with him in Turkey, when he and his team were caught in the crossfire of a country's internal rebellion, and it would probably be with him until the day he died. He called it his calm inside the storm.

"That night, the storm was blowing pretty hard. The captain, you see, he was a sort of a mysterious fella. He didn't ever really show up. That's to say, we all knew about him, but none of us had actually ever crossed his path, and we'd been out there on the same rig together, already for a few weeks."

Jim cleared his throat and glanced upward, smiling at the memory of the illusive captain. "That night we were busy securing the equipment on the deck. The rig was heaving, and work had come to a stop. The wind was wild. I'd never seen a storm so bad out on a rig. Once everything was secure, we headed back into the rig superintendent's office. Then suddenly, we see the shadow of this big guy filling the doorway. We knew the storm must be bad if he was showing up now."

Beryl jumped in wagging her finger. "And so you told him you needed a helicopter to fly out that very night."

"What??"

"You're kidding."

Jim smiled, mischievously, continuing with his story. "He had this manner about him, the captain. You see, he saw that we were concerned, and somehow he gave us confidence that we would get through it; that's all he did. And we believed him. And, of course, we were all fine."

"He inspired faith," Mona added insightfully.

"That's right." Jim leaned back in his seat, stretching out his legs, and then continued on. "The storm peaked at midnight. When the worst was over, I went to talk to the drilling superintendent and said that I needed to make a special trip out. He saw that I was determined about it—but not all hot-headed, impetuous-like. He knew that I knew what I was asking."

Beryl wrapped both arms around herself, shaking her head slowly.

"So, he agreed to it, and we requisitioned a helicopter. It flew in and hovered over the platform, using the landing pad as a point of reference because it was still too rough to touch down, you know, right on the rig."

Beryl grabbed her stomach, as if experiencing the seasickness from old, long-gone waves.

"And I held onto a rope mesh and crawled my way across the helicopter deck and up into the aircraft. And flew home."

Jim brushed his palms together twice.

"You're quite the daredevil," Mona said. Ted shook his head, chuckling, while Beryl added a footnote to her husband's story.

"I didn't know any of this, of course; he just showed up at home a day later with two first-class tickets to Oahu. He said he'd booked a stay at the Hotel Regent. I think he really thought I was planning to go to Hawaii with Tom Selleck!

Jim stuck his tongue out at his wife, who squinted, laughing.

"So, anyway, Aunty Vera convinced Uncle Carl to go too, and we all went to Hawaii together. It was fantastic!"

Ted laughed and in a good-natured way said, "You know what I call that? Stupid!"

Jim's eyes danced. He was hardly a daredevil; what he did know was how to put a situation into perspective, to factor out the static, set aside emotional content, see the big picture, and make the call. These were indispensable traits of a good leader.

As the girls' laughter trailed off, a more serious mood took over.

"You know," Jim began. "The fear was irrational. We knew every part of that rig. We knew what it could handle and what it couldn't."

"That's half the battle, right there, isn't it?" Ted drew in a quick breath and folded his hands behind his head.

Jim nodded, relaxing back into his seat. The girls exchanged knowing glances. For a moment, they enjoyed a comfortable silence, each in his or her own corner of contemplation.

Beryl was thinking of which vegetables to pluck from the garden for dinner, and Mona was wondering how she could help her. Meanwhile, Jim studied the backs of his hands, flipping through mental schematics of old rig designs, etched over thirty years ago.

As Ted sat there, his thoughts jumped back to the early days of Kumtor, before the crash, when they were still trying to win the contract with the government of Kyrgyzstan. The groundwork was done, but the politics of the new republic were muddling the situation, obstructing Ted's view and his one-page master plan.

Jim looked up from his hands and saw the distant look in Ted's face—when the mind flips a switch and reverts to mulling over old, unfinished business.

"You thinking about Kumtor again, or that fella there...from Peru?"

Beryl snapped out of her own reverie. Ted had already filled them in on that morning's disturbing news. She crinkled her nose, shuddering at the thought of someone severing his thumb and finger.

"Yuck...gross."

"He sure wasn't using his head." Mona flipped the front fringe of her hairdo with one finger and shifted in her chair. "Still—poor guy."

Beryl nodded, and then the girls leaned in toward one another to discuss dinner options. They were wearing matching long navy-and-white-striped tank dresses, which they'd picked up that afternoon from a sidewalk sale rack.

"OK, identical twin, what would you like for supper? We've got tenderloin or chicken, and there's kale and cucumbers and tomatoes from the garden. Peaches for dessert."

"You two look like the sisters in that Wrigley's Doublemint gum commercial," Jim teased, flirting with his wife.

Beryl put her arm around Mona and struck a cute pose.

"Mmm...chicken or tenderloin. Both sound good. Whichever, really. We're easy."

Then Jim turned back to Ted, who was distractedly following their conversation.

"Kumtor?"

Ted half nodded, chewing on a piece of cheese and cracker. Jim gave Ted his full attention.

"We'd been there and back twice, all the feasibility studies were complete and in order, but we'd yet to ink a deal. It was dragging on and on, and then the government received a nonconfidence vote from its own people for corruption, and right in the middle of it, the House was dissolved. We'd put in all this work but didn't know which way it was gonna go; the investment had already been significant. I was stressed and frustrated one day, and my colleague, John O'Hagen, just said to me, 'Ted, go down to the river and watch the seagulls shit.'"

Ted laughed as he recalled his associate's bizarre advice. "But I got the message and took it to heart, and did exactly that. It cleared my head, and then I could think straight again."

Jim nodded in agreement. "You know, Churchill went for ten-minute naps." He paused to form words for his next thought: "In a position of leadership, you have to have your wits about you, so you can see how she's going and where she needs to go—risk and reward, you know."

Ted, in his turn, nodded. He knew all too well.

In the end, with all of its tragedy and troubles, the Kumtor project had been a success and had catapulted Ted's career to new heights. But more importantly, it had made him a much better project manager. The girls exchanged looks, and then Beryl jumped up, turning to Jim.

"All I know is, I'm glad I didn't know the half of what you were doing out there in the middle of the ocean, or I would have had a heart attack!"

Everyone laughed, and she spun around and headed into the house. Mona stood up, and hearing the burden in her husband's voice, cautiously steered the conversation away from Kumtor, which could put her husband into a moribund mood. They were having such a lovely afternoon. She didn't want it weighed down by the past, with all of its heavy memories.

"That carpenter in Peru, he sure didn't weigh the risks with the rewards before reaching for that power tool."

"He certainly did not," Jim replied, "or if he did, his assessment was all skewed—screwy. That fella had too much confidence and not enough skill; he was cocky. That can cloud your judgment. Then you're heading for a fall."

Ted's eyes grew troubled. At Kumtor, clouds had literally obscured the pilot's view. The past was in the past, but the mind had a way of dredging up the details, like pieces of a bigger, yet unsolved puzzle. Answers didn't always bring peace. Fingers and thumbs may be reattached; physiotherapy, in some instances, could help restore original function and livelihood. But despite hope and best efforts, some losses were irretrievable.

Ted spoke, his voice half flat but exasperated: "What can you do? You give them a carrot but they don't always take it. At the end of the day, I can say, 'If you don't cut off your finger, I'll give you ten dollars.' But now, if you nick your finger, do you think you're gonna tell me? Hell no!"

Mona sighed quietly and considered joining Beryl inside, as much to preserve her own pleasant mood. She'd had enough shop talk.

"It's not enough for safety to be about good statistics," Jim began, offering a bird's-eye view. "No prevention plan is fool-proof, but still, improvements can be made." He paused, watching a sparrow that had just landed on the railing, close to them.

"You know, safety is not about individuals; you have to wrap the individual in the program, offer some kind of reward for working safely or pointing out safety issues, as opposed to throwing them a carrot after the fact for good statistics."

Ted lifted his head, perking up. That kind of talk he could understand. "You know, that sounds exactly like what Mark and I talked about this morning. We pretty much came to the conclusion that you can't reward individuals; you've got to reward teams."

"That's right." Jim lightly dropped his hand against the table.

The bird flew off, retreating to a nearby tree.

Ted brainstormed out loud. "Whether it's monthly, annually, or on a project basis. Maybe it's a baseball cap, or a night at the airport pub when they fly out—something symbolic that rewards everyone for the safety of the entire group." Jim jumped in, summarizing the ideas.

"There are guarantees, but the focus shifts. All of a sudden, people want to work safely for the strength and welfare of the whole team. And then, what happens? Participation skyrockets; they have fun with it; it builds morale. I think you fellas are onto something there."

Ted's face relaxed, and he took a deep breath and then a sip of cool white wine. Had his cousin-in-law worked in the same field as him, he would have wanted him on his team. They shared the same quality standards and overall vision.

"Good," Mona said half teasing. "Now can we talk about something else besides work?"

The guys looked quizzically at each other. The line between work and life could barely be distinguished some days. Just then, Beryl emerged from the house with a tray of appetizers and noticed the energy had shifted.

"What are you guys talking about now? This is supposed to be happy hour!" She shook her hips until everyone finally cracked a smile.

Ted looked apologetically at his wife and puckered his lips in a kiss. Then he smiled at Beryl and Jim, grateful for the shared space of friends. He was coming to believe that who you had on your team made all of the difference. Projects were living organisms. People were key, their skills, talents, and integrity paramount. Morale and safety hinged on knowing your team and trusting them—even with your life. Ted sighed, and smiled. Life was good, and these people in it had his back. Some days, that was as sure as life got. You just never knew what news you might get when you picked up the phone.

Ted raised his glass. "To happy hour!" And they all joined in.

They toasted four or five times—to family, fine wine, Okanagan sunshine, ever-patient wives, and the perpetual limbo of Jim and Ted's slow, steady dance toward something possibly resembling retirement.

One day. Not quite yet.

Finding Balance

LESSONS LEARNED
1. It's easy in the heat of battle on a project to forget your commitments to your family. Good project managers find a way to balance work and home life.
2. Good project managers inspire confidence.
3. International project work inevitably demands extraordinary commitment and compromises within the family.

CHAPTER THIRTEEN

Siberia

After Bre-X and its disappointments, Ted could have used a good project to bounce back. Instead, he got Pan American Silver and the Dukat Mine in Siberia.

Russia's largest silver deposit, located in the far eastern Magaden region, had great potential to be one of the world's most abundant lower-cost silver mines. The mine was shut down in 1995 due to inadequate state funding during hard times in the new post-Communist nation. After sitting idle for a couple of years, the government decided to tap the wealth of its national treasure and invited in foreign investment to help rebuild. Pan American entered the picture, scooping up a 70 percent stake in the mine, and then contracted global conglomerate SNC-Lavalin to manage the project.

Ted was sent overseas with his best team to run the show. He was happy enough to go, having wanted a change of scene ever since he'd grown disillusioned by the head office and his company's current leadership. So he'd asked for a package or a project. They gave him Siberia. Russia. Why not? Ted had thought.

Mona had packed his fur hat with earflaps but opted to stay at home in the relative warmth of a Toronto winter.

And so, with renewed optimism, Ted had embarked on another foreign adventure. The feasibility study was already complete, and on-paper projections for the silver mine looked promising. After arriving on Russian soil, however, Ted and his trusted engineering team paid a visit to the Dukat site, only to see the sorry state of the whole mining operation.

***

"There were no systems in place. It was a total gong show!" Ted exclaimed to Jim, on a lazy Wednesday afternoon.

The two were back at it, lounging on the Cooke's deck, trading old work stories like the vintage baseball cards of their youth. Engrossed in Ted's story, Jim listened, while doodling with a mechanical pencil on a pad of graph paper.

That morning the guys had played a round of golf with their wives at a nearby course, and now the girls were inside sipping iced tea in the welcome respite of the air-conditioned living room, while the men retreated to the shady veranda. Had either of the girls been listening to their husbands' conversation that afternoon, they would have found it as dry as a piece of unbuttered toast.

But Jim hung on his cousin-in-law's every word, as Ted swirled the ice in his glass, recollecting the organizational mayhem at Dukat.

"We had already done a feasibility study. They had the project definition right; we had a place to start from—the potential was there—but there were no control systems in place. They were nonexistent!"

Jim nodded, intrigued, as only a fellow engineer could be. He loved nothing more than a good quandary. "So what did you do?" Jim probed.

Stretching out on his lounge chair, Ted replied matter-of-factly. "We started from scratch. We invented our own cost, scheduling and reporting procedures, and we tried to bring the workforce and staff on board. We even set up simple accounting for the owner to pay the bills and track money flow."

"Wow, they were really, in a poor way," Jim said, turning back to his notepad.

Ted nodded, folding his hands behind his head and collecting his thoughts. He was enjoying the cool breeze of the nearby fan. Ted felt more relaxed than he had in years. He'd even shot an 89 that morning. It seemed that their little holiday and all the fresh air, good wine, and company had improved his handicap that week. Maybe there was life after project management, after all?

Ted contentedly scanned the Cooke's vast backyard full of trees, food gardens, water features, and flowerbeds. When Jim wasn't busy implementing his latest home project, he was quietly planning the next three on a trusty pad of graph paper. The mechanical pencil placed in Jim's strong, steady hand was like a bridge between man and world. Given a notepad, most people would thoughtlessly doodle curlicues or flowers or tic-tac-toe grids. However, as Ted and Jim had been shooting the breeze that afternoon, Jim had idly drawn a graph of the law of diminishing returns.

He came by it honestly. Just the other day, much to his family's amusement, he'd pulled out a notepad over cookies and tea to try to explain to his teenaged grandson why professional baseball teams hired new star players to increase their winnings but how, at a certain point, signing on more star players paid off less and less, so was no longer worth it. A simple grid and line graph did the trick. Jim lived to see that light go on in his grandkids' eyes when he was able to explain how something worked. Earlier that spring, he'd even drawn a bar graph for Beryl as they discussed the best varietals of tomatoes to plant for that summer's garden. Beryl had looked at the columns and then looked at Jim and burst out laughing. It left him confused and giddy, just like on their first date, in 1959, when watching a movie she'd leaned in and asked Jim, "So what do you do for fun?" And he replied, simply, "I achieve something." Jim's idiosyncrasies were a running joke in the family.

An engineer's mind is a strange and mysterious thing. Jim was a wizard when it came to winches, pulleys, circuits, gizmos, and gadgets; he had systems rigged up all over the property to maximize efficiencies, streamline tasks, and organize space, bringing logic and order to their cherished home, which he'd also conceptually designed from the ground up. These were his hobbies, but in his project career, four management principles led his everyday thinking. He'd learned early that even though his work was mainly technical in nature, the principles of planning, organizing, leading, and controlling were critical for successful completion of a project. If these principles were not understood in an organization, or by its project manager, Jim would teach the key people the meaning and importance of the principles before the project began and guide them personally throughout the project itself. This was probably the most difficult part of Jim's career since it took a lot of time and attention, but it was always the most rewarding and, at the end of the project, gained him considerable respect. He knew these four principles reduced project risk significantly and were largely the keys to successful completion.

Ted, on the other hand, was the first to admit that he was a terrible engineer. If anyone asked him to design something, he liked to joke that he "wouldn't stand on it." But Ted knew where to find the guy who could do the job for them. This made him an invaluable project manager. He had the contacts and the connections. People liked and trusted him. He wasn't afraid to roll up his sleeves and delegate the dirty work.

A few minutes passed without notice, and then Ted and Jim seamlessly picked up their conversation.

"It sounds like you had the right core team—that wasn't the problem," Jim recapitulated, setting down his pencil. He crossed his ankles and took to examining his running shoes.

"No. I'd handpicked them myself. Everyone knew how to work together and knew what needed to be done. With a management team like that, you can cut six months off of your schedule."

Jim perked up like Ted had just rattled off some impressive Major League batting-average statistics.

Ted casually glanced over at Jim's pad of paper. "What are you doodling there?"

"Oh, that's just the diminishing returns line. You know it."

"Yeah, sure." Ted nodded unfazed.

Through the glass door, Jim could see the girls sitting at the small media perch he'd built onto the kitchen counter. They were looking at photos on Beryl's Facebook page. He didn't get that whole social media thing. Then, still cogitating, Jim turned back to Ted.

"You know, these days, with oil supplies dwindling, and what is left is much more difficult and more technically focused and expensive to get out of the ground"—he rubbed his thumb and index finger together—"every project has its own price tag."

"Sure," Ted agreed. "It's gotta be worth it. It needs to show a reasonable rate of return." Whether the resource was silver, gold, oil, or gas didn't matter; a project had to be worth the money and effort invested. Too much or too little of any number of factors could tip the scales and turn a project's viability into a liability.

"The cost-benefit has to be there; if not, somewhere along the way, you're gonna lose."

Ted summed up the situation at Russia's silver mine. "The problem was they didn't have the operations expertise or the facilities in place."

"Or the mind-set, sounds like," Jim threw in, without looking up from his shoes.

"Or the mind-set," Ted echoed. "The Russians simply had no sophistication when it came to major project development." He lingered a moment on the thought. "Good people, great vodka—but piss-poor projects. Back then, anyway. Times may have changed."

"I've seen it before," Jim began, "in Communist Ukraine."

"Is that right?" Ted crossed his ankles, relaxing deeper into the lounger.

Rubbing his chin thoughtfully, Jim recalled a near-forgotten memory of a work trip long ago. "We visited an oil field at the foot of the Carpathian Mountains," he began. "It had the potential to be a first-class project, but it was a third-rate operation—just like your Dukat."

Ted gave Jim his undivided attention.

"It was back in the 1990s, I was going there to evaluate a field with the thought of an acquisition. We traveled from Frankfurt to Kiev on a Russian airline, and they'd used duct tape for maintenance on their plane."

"Uh-oh," Ted laughed. "That was the first red flag."

"You're telling me." Jim paused to recollect the next leg of the journey. "We took a train from Kiev to L'viv and traveled in a Volkswagen type of a van south to the oilfields where we visited one of the geologists. He was a nice guy, highly educated, but he made only one hundred dollars a month."

"No wonder their cars looked so rough."

Jim inched forward in his seat. "You know, Ted, I entered the Ukraine with a briefcase full of maps and confidential information. I don't know how I entered and left without being pulled over by the police. They must have known who I was and why I was there and cleared the path in both directions. I didn't even go through immigration."

Ted shook his head. "You're kidding. Sounds like they were paving the way, doing everything they could to woo the business."

Jim sat back, resting his arms across his chest. "That's exactly right."

Then Jim chuckled, his shoulders bouncing slightly.

"What?" The corners of Ted's mouth turned up.

"I remember, the day we arrived, our driver took us on a tour. We were in a Volkswagen with no side doors to get in and out. This fella had brought his secretary along, who'd packed a big picnic lunch, complete with blankets and cold foods, you know, to showcase their local cuisine—and of course, several bottles of vodka were packed in the basket."

Jim raised one eye, leaning in even closer. "You know, at a dollar a bottle, they took the stuff everywhere they went."

"Sounds like Siberia...and Kyrgyzstan." Ted laughed, thinking of his old pal, Duchaine Kassanov, the hockey fanatic. "Go on."

Jim cleared his throat. "Anyway, after an hour of driving—it was a bumpy ride—we stopped, and his secretary set up a blanket on the side of a hill. We sat down, and she uncovered all of this food—a big spread—and I couldn't tell the difference between what was what. So I grabbed a square chunk of what I thought was cheese." Jim cringed, his mouth twisting at the unsavory memory.

"It wasn't cheese?"

"No!" His eyes widened. "It was pork fat, a local delicacy. But, it was the roughest, most horrible stuff."

"Did you spit it out?"

"No, I washed it down with a swig of vodka, and they thought I liked it, so they offered me more."

Ted snickered.

"Anyway, they were extremely hospitable and really, highly educated people," Jim continued. "Individually, they made a good impression on me."

"But they just didn't have the 'field,'" Ted said, jumping in.

"They did not." Jim's hand dropped to his leg. "It was dirty and old and improperly maintained—and would have opened us up to all kinds of safety hazards."

"Hmmm," Ted nodded. "I hear ya."

"Not to mention, their wages were so low there was little incentive for the workers to perform. I remember the geologist's textbooks were even turning yellow with age." Jim paused, looking at Ted. "That's Communism for you."

Ted nodded with resignation.

"It was too bad. The oil was there—but the operating philosophy was not. It would have been too difficult to manage. We couldn't touch it."

"The reality of the situation is they weren't quite ready to consider performance, you know, play in the big leagues." Ted crossed his arms.

"Sounds like your project." Jim passed the torch. "So what ended up happening in Siberia? Did you pack 'er in?"

Ted raised his glass, and then realized it was empty. Jim jumped to his service.

"What'll you have, Ted?"

"How about...hey, a vodka martini?"

"Good call. I think I just might have one too—if Beryl lets me." Jim winked, glancing inside at his wife.

"Actually—we got sent home."

"What?"

"Yep."

Jim scratched his head. "What do you mean you got sent home? From Russia?"

"We got the ball rolling for them, got operations in place, trained their people—we were building the project—but then found out that Pan American still had to bid on the rights to the mine. They went about it backward, started the project before they had this approval. We were proceeding as if we already had it. Then they lost the bid. I'll tell you, we learned a hard lesson: nothing is certain until all approvals and financing are in place."

Jim shook his head. "Amateurs."

"That's right. The guys knew nothing about international business and projects."

"You don't want to deal with them."

Ted and Jim looked at each other in mild disgust, and then they looked at their empty glasses. Then they looked inside to where their wives were hanging out in the kitchen.

Jim rubbed his calves, and Ted stood up and stretched his long legs.

"So it was a lemon?"

"It was a lemon."

"What can you do?"

Ted shrugged. "Try to make lemonade? And if it doesn't work, I guess, cut your losses and get the hell out."

Jim agreed. "That's all you can do."

For Ted, not all had been lost in Siberia. He'd learned a valuable lesson from yet another project failure that would hopefully one day net him another success. He was long overdue for one of those. Russia had taught Ted that good people will make a bad system work. But a good system, once in place, won't save a bad project. Nothing is certain, and furthermore, if you are not enjoying what you are doing, then you must make a change, no matter how difficult it is.

LESSONS LEARNED
1. If you are not enjoying what you are doing, you must change no matter how difficult it is.
2. Nothing except feasibility studies should proceed until all approvals and financing are in place.
3. Bad politics and limited experience will kill a project.
4. Every project has its own unique circumstances that can range from political and environmental to economic and technical matters.

CHAPTER FOURTEEN

Diavik

Ted finally got his "perfect" project. Just one day after he'd returned from the cancelled job in Magadan, Russia, his boss assigned him to Diavik, a mine site located on Lac de Gras in the Northwest Territories, approximately three hundred kilometers northeast of Yellowknife.

It was December 17, 1999. Diavik Diamond Mines. Inc. (DDMI)—the manager of the whole project—had hired SNC-Lavalin on a joint venture with the Dobrib-owned Nishi-Khon to do the engineering design and procurement on Diavik. Another company, H.A. Simons, had been brought on to do the construction management. There was a twist to Ted's assignment.

"We need you out in our Vancouver office, where there will be more staff and mining project expertise available to you," said Roger Nichol that day.

"OK, but what about Harry?"

Harry Sambells, a manager in their Calgary office, had completed the project feasibility study. Typically, if you did one, you did the other. But apparently something else was driving the decisions over at head office.

"He just doesn't have the construction project experience we need," Roger explained.

"Construction experience? This is E&P. What does he need construction experience for?" Ted was confused.

"Ted, we want you to steal the construction contract from H.A. Simons."

"OK, now I get it."

Ted had what Harry lacked in spades. He was their guy for the job.

So Ted and Mona packed their bags on January 3, 2000, leaving their Oakville, Ontario, house in the care of their son, Paul, and joined their daughter, Anita, who'd recently relocated from Prince George to Vancouver. They found an apartment in Kitsilano, known for its great beaches, and settled in nicely for the duration of Ted's assignment, which turned out to be eighteen months.

***

"The timing couldn't have been better. We got to spend a lot of it with Anita and Mike and got to know our future son-in-law. Those were great times," Ted reminisced, as he and Jim strolled through the east Kelowna neighborhood. It was almost happy hour again.

"So how did it all go down?" Ted's cousin-in-law asked, as he simultaneously examined the pleasing design of one neighbor's yard, its focal point birdbath and abundant flowerbeds.

"That sure looks sharp, doesn't it?" He pointed to a robust rose bush and hydrangeas in the center of the yard.

"It does..." Ted said distractedly. He could hardly wait to tell his story.

Jim turned to his cousin-in-law, giving him his full attention. "So, did you nab the contract?"

Ted beamed with unmistakable pride. "Sure did!"

"Nice!" Jim clapped his hands, frightening a nearby squirrel up a giant oak tree, and chuckled. "How did you do it?"

"Well, first of all, you see, DDMI, the manager, had engaged Philip du Toit to act as the VP of projects."

Jim scratched his head. "I've heard of that guy. He's with the cream of the crop."

Phil du Toit was an accomplished mining executive with over thirty years executing large capital jobs for some of the world's leading mining companies.

"You don't step on his toes, that's for sure. And that's exactly how the H.A. Simons guys tripped up."

"Oh?" Jim continued walking, and Ted followed, talking has he went.

"My strategy was to deliver on time with minimum disruption to DDMI. Within three months, the objective was achieved, and H.A. Simons was removed from the project."

"There's no arguing with results," Jim said, looking at a small bird on a nearby fence as they passed.

Ted turned to Jim. "Sure, proven performance is one thing. But knowing people and how to deal correctly with the various ranks is another matter."

"True enough." Jim nodded. "The other guy didn't get that."

"He did not," Ted explained. "Part of the reason H.A. Simons got cut was their manager acted as though he was boss of the whole operation. He usurped the authority of Phillip du Toit. He didn't know his place."

"You've gotta know your place." Jim shook his head, stopping as they arrived at the curb, a block from home.

At the crossroads between people and performance, Ted had achieved his mandate with flying colors. He'd nailed the job. By June of 2001, the engineering component of the project was complete. Ted and Mona had given up their apartment in Kitsilano and relocated back to Oakville, Ontario. Then Ted moved up north to the site to manage mine construction.

Jim veered left, and the men crossed the street.

"So, you finally got your perfect project."

Ted let out a sigh of satisfaction. "Diavik was my ace in the hole, that's for sure. The project came in under budget and ahead of schedule. The scope was crystal clear, and there were no significant changes to project scope once detailed engineering began."

Ted's arms swung confidently by his side as he headed back toward the house. The sun was making its way back toward the horizon. Ted felt a hunger pang as he smelled backyard barbecues being fired up.

By the fall of 2001, project construction was moving along smoothly; so smoothly, in fact, that Ted grew restless. He saw no new challenges up ahead, and one day the barometer had read minus 77 degrees Celsius. So Ted talked to his boss, but all Roger would offer him was a business development position, which didn't excite him. So Ted resigned his position with SNC-Lavalin and, once again, opened the doors to chance and a new adventure.

"But speaking of people, part of the reason my work at the mine site was so easy was that I had my associate, Nick Mills, there with me, working as the construction manager," Ted added, matter-of-factly.

"The same guy you had on Kumtor?"

"That's right. I could always count on him to do a good job. And after that early change in construction management, there were not any significant personnel changes on Diavik."

"Change is good until it's bad," Jim threw in, philosophically.

He was getting hungry also and was looking forward to kicking back with a rum and Diet Coke before dinner. Still, chewing on one of Ted's great stories was as satisfying as sinking his teeth into a steak, or the seventh inning of a head-to-head Red Sox game. He straightened the cap on his head, looking at his cousin-in-law expectantly.

"Management was willing to go the extra mile and provide approvals in a timely manner. So there were no delays to project execution, and Phillip du Toit was a tough but fair client who demanded performance but knew what was important to make projects work."

"There were no major foul-ups."

"Well, not exactly." Not everything on Diavik had gone exactly to plan. "It was a perfect project, except for the fact that we had two fatalities on that job—a safety issue."

"What happened?"

Ted told the story as they stood at the end of the Cookes' driveway. One evening, around 11:00 p.m., after Ted had gone to sleep, there was a knock at his door. "It was our team's mechanical superintendent, Dennis Krahn, and he told me there had been an accident."

Ted jumped into some clothes and Dennis had driven him to site. "That night, two workers were attaching cladding to the process plant when their man lift toppled." Gregory Wheeler and Gerhard Bender had come crashing to their deaths.

Jim shuddered. He'd been on projects before where similar accidents had happened. "Was the structure compromised or did they fail to follow safety regulations?"

Ted rubbed his forehead, other hand on his hip. "Actually the investigation revealed that the lift had been inspected before being brought to site. It was structurally sound, but the electrical switches that controlled the lift's movement had been overlooked, and they were corroded."

Jim shook his head. "If it's not one thing, it's always something else."

Ted nodded. "It was a costly mistake. One that cost two men their lives."

"A terrible tragedy. So not a 'perfect' project, after all."

Ted rubbed his chin. "No, I guess not. Still, Diavik was a stellar success, by every other performance measure."

LESSONS LEARNED
1. Some projects have it all – defined scope, a realistic schedule and cost estimate, and a great team.
2. Thorough equipment check-outs are mandatory prior to starting work.

CHAPTER FIFTEEN

Great Barrier Grief

Ted's next project found him in a hardware store. That weekend Mona had asked him to fix a leaky eaves trough before winter came, so he'd thrown on his baseball hat and headed over to the local Home Hardware.

He was strolling through the aisles in a plaid shirt and his weekend jeans, tossing putty knives, caulking guns, and roof cement into his basket, when he heard a familiar voice call his name. He turned to see Russ Buckland, an executive headhunter whom he'd come to know quite well over the years. Ted waved, and Russ walked over.

"Hey, Ted. How are things?"

They shook hands like old friends.

"Good, good. And you?"

The two men bantered for a couple of minutes about home renos and summer holidays, and then the conversation shifted to work.

"So I hear you left SNC-Lavalin and you're looking around," Russ said. It was his business to know the scoop on every projects executive around.

After Russia, Ted and his team had gotten their success. The Diavik Diamond Mine up in the Northwest Territories was a "magical" project, coming in ahead of schedule and under budget. It was so successful, in fact, that Ted had grown bored and decided to leave.

"I've got a job for you," Russ said confidently, without sounding pushy.

"Oh?" Ted adjusted his glasses. "What is it?"

Ted liked the man, and he trusted his instincts. On several occasions, over the years, they'd negotiated on potential contracts. Russ understood the business and didn't peddle jobs like a used-car salesman. He gave the facts, built strategic relationships, and let the chips fall where they may. A few summers back, they'd socialized, playing a few rounds of golf with their wives and having dinner and drinks. Although Ted didn't need work, he was starting to get the itch.

"Inco's looking for a director of projects to oversee their capital projects. The initial assignment will be to negotiate business agreements with Innu and Inuit up in Labrador for the construction phase of the Voisey's Bay Project."

"Is that right? I heard there was some struggle up there with negotiating stake holdings and building consensus on a number of issues."

"You got it."

Ted set down his merchandise basket a moment and crossed his arms. "Sounds interesting. Why don't you send me the profile and I'll have a look. I'm about ready to sink my teeth into the right job."

"Sure, Ted. I'll do that," Russ smiled warmly. "Well, I'd better finish up here. I've got some flooring to lay. And hey, great to run into you. Say hi to Mona."

The two parted ways down aisle six. After grabbing some silicone and an extra roll of duct tape for good measure—you could never have too much emergency glue—Ted headed to the tills.

Nice guy, Ted thought, as he paid for his supplies and walked at a half clip out into the sunshine and toward his BMW. He felt energized by their chance meeting. It was time for a change and working for a large Canadian outfit like Inco was a logical next move in his career.

The mining industry had been hit hard by the global financial crisis in 1997. Investors had lost billions, and the Bre-X scandal had caused the departure of record numbers of professionals from the industry. They were now in a bear market; commodity prices had dropped, pessimism was high, and smaller mining outfits were facing increased challenges accessing funds to finance new projects.

Inco was one of the biggest, most prosperous companies in Canada, and the largest nickel producer in the world. The work sounded complex and high level. Ted's remote mine project experience over the past three decades would be a definitely asset to such a project. Voisey sounded like a natural fit. Besides, Ted was done with overseas work. He was ready to stay close to home; this contract was based in Ontario, although there would, no doubt, be a bit of northern travel.

He looked forward to reviewing Russ's summary.

Adjusting his hat, Ted headed home to tackle his home repair project. He was no Mr. Fix-It, but by then, he was well known in his field as a handyman in his own right; a top-notch project director with a long career of reputable work in his proverbial tool belt. How hard could it be to repair a leaky eaves trough?

***

Ted had been working on the Voisey's Bay Aboriginal agreements for three months when he got an unexpected call from his new boss.

"Ted, how's the IBA going?" asked Terry Owen, Inco's VP of projects.

Sitting before him, Ted had the latest marked-up draft of the impacts and benefits agreement. Times were changing. New dimensions and challenges were arising in project development work due to the new strong emphasis being placed on social and environmental sustainability issues. The IBA was a legal contract, increasingly being used in the industry to formalize relationships between mining companies and aboriginal communities, whenever proposed projects fell on First Nations lands.

Ted was satisfied with that morning's progress. The Newfoundland and Labrador governments stood to gain some $20.7 billion of their GDP from Voisey's Bay mine over the next thirty years, and so they were very amenable to negotiations with their local Innu and Inuit communities.

"I think we've finally come up with a list of terms that both parties can agree on."

"Good, good."

"It's coming along well, Terry."

Ted leaned back in his ergonomic leather chair, scanning the neat, detailed margin notes in his Day-Timer. This idiosyncratic logging routine had served him well over the years, especially during the Kumtor crash inquiry. If he ever had to recall the specific action steps, events, or key conversations of a given day, he need only look up the week, month, and year. His fastidious note keeping went back to 1976. He'd never give it up. Ted gave Terry a short, concise summary of that morning's progress on Voisey.

The nickel-copper-cobalt deposit, located off Canada's north Atlantic Coast, was considered to be one of the country's largest mineral discoveries over the last four decades. Inco had paid a whopping $4.3 billion in 1996 for rights to mine 140 million tons of sulfide ore containing nickel. The close proximity of local Innu and Inuit communities to the mine site was raising concerns, so they were being proactive to keep the project moving forward on target. Therefore the goal with the IBA was to reach consensus on terms of economic benefit for the First Nations groups and build an approved plan to minimize and manage the social, cultural, and environmental impacts on local communities due to mining activity in the area.

"It's all coming together—government royalties, Innu and Inuit employment and contractor rights—we're almost there," Ted concluded.

"Super. Good work," Terry replied, on the other end of the line. "Say, Ted, have you ever been Down Under?"

"Pardon?" Ted laughed. The question caught him off guard.

"We're having some problems on Goro Nickel, over in New Caledonia."

"Oh?"

"I've got to get down there as soon as possible to find out what the heck's going on. Can you join me for a week, say, in early January?" Terry asked.

"You want me in Australia?" Ted was baffled.

"Well, sounds like you've got Voisey under control for now."

"Well, yeah, but..."

"We could really use your input down there, Ted." Terry's voice sounded panicked.

"But, Terry, why? I mean, isn't Ron Shepherd taking care of things on Goro?"

His boss sighed. "Ted, I honestly don't know. He's been down there for three months, and the schedule's already a mess, the budget's ballooning, and whenever I talk to Ron, it sounds like he's barely keeping his head above water."

Goro was a big-time project, and Terry's ass was on the line.

"We've got to get things back on track. This is too big to foul up."

"OK...sure, no problem, Terry. Send me the dates. But, you know, Terry, I'm finished with overseas work, right?"

There was a split-second delay on the other end. "Yeah, I know, I know, Ted. Don't worry. It's just a few days. Then you're back on Voisey. I promise. But I could really use your input on this." Terry was virtually pleading with him.

"Well, then I'd better brush up on my Aussie."

"Super, mate."

"And, I guess, my Français. Isn't New Caledonia a French colony?"

"Uh...Oui, oui."

Terry's voice grew calmer, but just beneath the surface Ted sensed desperation in his boss. What was he wading into?

"Just a week. Thanks, Ted. Right-o."

***

Ted hung up and then dialed a few of his former colleagues who would know the ins and outs of the Goro project. He wanted to pick their brains a bit. His old cronies filled him in.

Goro was one of the largest nickel laterite deposits in the world, low grade but high volume, with cost estimates of $1.45 billion and big-time potential payoffs in the hundreds of billions. The mine site was located just off the east coast of Australia in the colony of New Caledonia, an island utopia, conquered by the French in the late eighteenth century. The land was rich in culture, ecologically diverse, and full of nickel ore.

Ted didn't want Goro. He didn't want to travel halfway around the world to work. That year he'd turned fifty-five. His daughter had gotten married to Mike. His son had moved back home. Mona wanted him around. It was time to stay local and let the sophomore engineers cut their teeth on the international projects.

His brain told him no; still, he felt a pull. He'd always wanted to visit the South Pacific. Besides the quick trip would be a nice break from the Voisey's Bay contract.

But something was nagging Ted. Why the urgency in Terry's voice? What kind of trouble could there be at Goro, so early on in the project, to warrant a trip by both the VP and director of projects? His boss had handpicked the technical and management teams he'd sent over, and they were already set up, presumably getting mine operations underway. It didn't make a whole lot of sense. Ted figured he'd see soon enough.

He was pleased to have his bosses' confidence in his senior capabilities. Terry was an affable if sometimes short-fused guy who would be an enjoyable travel companion. The Aussies, along with their French neighbors, not only had their share of golf courses but arguably some of the finest wines in the world. Maybe there would be time to work in a game. Mona might be a tad envious, but given the whirlwind trip she wouldn't be able to accompany him. Not this time, anyway. Maybe for their thirty-fifth wedding anniversary the following spring they could take a golf and wine tour Down Under.

Ted speed-dialed his wife.

"Hi, Ted. What's up?"

"Not sure what's 'up' but I know what's going 'down.'"

***

Brisbane was a broiler. It was 33 degrees Celsius outside in January, at the peak of Australia's summer. The cool, climate-controlled conference room of

Great Barrier Grief

Goro Nickel's management offices was no welcome reprieve.

Ted couldn't get over the bunch sitting around the table. Goro's project team looked like they'd rather be chased by a herd of camel in the Outback than spend a minute more in that room together. The absence of idle chitchat or even work-related banter was not a good sign. What weren't they saying? Ted wondered, as he pulled up a seat.

One thing was clear. This team had no gel; that stick-fast quality that binds a group together by a common credo or purpose. It was glaringly missing. Something was wrong. And they were there to get to the bottom of it.

"OK, let's hear it, Ron," Terry leaned back in his chair, crossing his arms across his chest. Ted straightened the legal pad in front of him and then turned his gaze toward Inco's project director.

Ron made a few opening remarks. Over the next ten minutes, Ron gave them a summary of progress on mine development to date, and then turned the floor over to Bal Panicker to lead the engineering contractor's presentation. Next up, Bal gave a brisk rundown of recent changes to the project schedule, a few cost issues, and unresolved contractual problems that were contributing to delays.

No wonder Terry was worried. It was basic project management protocol to ensure that all key contracts were signed and in place before a project got underway. With billions of dollars on the line, what were these guys thinking, moving ahead and spending money, before their ducks were in a row? Especially since the engineering contractor team was on a target price contract for the capital costs schedule.

Terry pursed his lips and said nothing. Instead he motioned with his hand for Technip's construction manager to give his update. Clearing his throat nervously, the French engineer began to sweat bullets as he detailed why, six months into the schedule, mine construction had not yet begun.

This was getting worse and worse. Ted couldn't believe his ears. Apparently, many of the process design decisions had not been made, including the fact that test work was still underway on a state-of-the-art acid leaching system they'd licensed from a South African outfit, which would optimize their production.

James Marcuza, the engineering manager for Hatch, spoke up. "As we speak, a new pulse column technology is being evaluated for its safety and benefits over traditional mixer settlers. Yes, pilot-plant operations are, indeed, behind schedule, but the important work being done there will determine the most cost effective and least invasive way to extract nickel," the engineer continued.

Terry gruffly cut in. "Didn't they figure all this out already?"

"Yes, but it's not that simple." The Hatch engineering manager spoke deliberately. "Nothing like this has been attempted before on such a large scale. This is industry-changing stuff here. We're charting the future course of solvent extraction."

Those days, they had to process increasingly complex raw materials with often decreasing grades of metal—all on top of the fact that mining operations were facing stricter limits on their allowable impact to the environment. The industry was increasingly stuck between a rock and a hard place.

"Yes, yes, I know." Terry tapped his fingers nervously on the desk, looking like he was about to come unhinged. Then he glanced sideways to Ted and back to the crew. Everyone waited for his reaction.

Terry cleared his throat. "So what I'm hearing"—his voice increased several decibels—"is that we are behind schedule, off budget, unclear on several key areas of project definition, technologies, and time lines, and we haven't even broken ground yet." Terry stopped to glare at everyone around the table. "Not to mention, the systems we need in place to get the ore out of the damn rock aren't even tested and ready to go! Jeez, Louise, who the hell is running this damn show?"

Good question, Ted thought. It was clear to him that no one in the room, including his boss, had a handle on the project. The scope was muddled, direction uncertain, and leadership lacking. Goro was in big trouble.

Terry stacked the papers in front of him and pushed his chair back from the table. "I've heard enough."

Ted followed his lead, closing his pad, and lining up his pens.

"We'll pick it up tomorrow, nine o'clock," he said, in clipped tone, and then turned to Ted, in stunned silence.

Everyone filed quickly out of the room, trying to make themselves as inconspicuous as possible. Terry looked seriously stressed. Ted didn't envy his position; then again, his boss hadn't brought him in simply to watch the whole debacle.

"What do you say, mate?"

Ted ran his hand threw his hair, unsure of where to start. "What a mess."

"You're telling me. Can we talk somewhere privately?"

***

An hour later, Ted and Terry were sitting on a shaded patio. Ted was enjoying a cold schooner of XXXX, a Queensland beer. Terry sipped a water. He didn't drink alcohol. Perspiration lined his upper lip. They looked at each other. Terry spoke first.

"Ted, I knew it was bad, but this is much worse than I thought. They're like chickens with their heads cut off. No one knows what the heck is what. It's a total shit show!"

For several minutes, Terry railed on. Ted listened, letting him get it out of his system. When Terry finally exhaled and sat back, Ted offered his take on things, beginning with something positive.

"That Hatch engineering guy, he seemed to know what he was talking about the most. It's exciting stuff that they're working on, really innovative."

"Yeah, but"—Terry looked impatiently at him—"I'm not sure Goro has the scope and budget for these kinds of delays. Not now, with today's markets and costs of doing business. At this rate, they're going nowhere fast."

Terry sat back in his seat, relieved to hear that Ted had wrapped his head around the issues. His own head was spinning.

"What are we going to do? Those two are gonna kill each other," Terry said.

Ted didn't like how his boss had said "we." Ted had worked in the business long enough to know when people were the problem.

"I don't know what the deal is between Ron and Bal, but together they are sending this project south, and fast. Who's in charge, anyway?"

Terry looked blankly at Ted.

"They obviously haven't had their responsibilities defined. They're stepping all over each other."

Then Ted stopped, deciding to tread carefully. After all, Terry had hired Ron and approved the tripartite venture between Bechtel, Technip, and Hatch. He'd built this team and set them loose; whatever problems arose, as VP of projects, the onus for straightening them out sat squarely on his boss's shoulders.

Then why did Ted feel like the yoke was being passed off onto him? Ted could sniff out a management problem in an airtight container a mile away. He knew how to pull the wheat from the chaff. It must have been the Saskatchewan in him. That made him indispensable. Terry knew it.

What had Ted gotten himself into?

* * *

The next morning was repeat boardroom bedlam. Ron and Bal were at each other's throats again.

At lunchtime, Ted made a beeline for the cafeteria. The white noise of the sanitized space soothed his nerves. He emptied his mind as he scarfed down an order of Aussie fish and fries, some coffee, and mango-lychee pudding. Feeling his legs beneath him again, he sighed with satisfaction. Thankfully, he would not have to see the inside of a boardroom for the rest of the day.

By midafternoon, Ted was flying high across the glistening South Pacific in a chartered plane, on his way to the Goro mine site on the island of New Caledonia. In his company, for this little field trip, were Terry and his old friend and colleague Andrew Greig, president of Bechtel's Mining and Metals Global Division. The change of scene lifted everyone's mood.

Looking out across the water, Ted grinned. "What a day. Look at that. Spectacular!" he pointed at the green archipelago in the distance.

"Tough day at the office, men?" Terry rejoined, cheek pressed to window. "But well-deserved, don't you think?"

Andy chuckled, relaxing deeper into his seat. "You couldn't ask for a more stunning work site."

"I'll say," Ted added.

No one said so, but they all hoped that setting foot on solid ground at Goro would help bring much needed clarity and direction to the project. Getting out of the office was a bonus. No PowerPoint or projection charts in their immediate future made for bluer skies and sunshine all around.

Goro was every mining company's Shangri-La. New Caledonia contained an estimated one-fifth of the world's nickel resources, all on an island oasis of just 18,850 square kilometers. A lush, tropical island located 1,500 kilometers east of Australia, she was home to a unique and diversified ecosystem, including the world's largest lagoon and the second-largest barrier reef. The island people consisted of about 250,000 European and mostly French settlers and indigenous Kanak clan members, who were exceedingly proud and protective of their abounding paradise, and for good reason. The extraction of minerals on the island in the past had come at a cost to the people and environment.

Ted had done his homework. Even so, he asked Andy to brief them on their way over to the Bureau de Recherches Géologiques et Minières, where they were meeting a government team for their tour.

"Goro Nickel is the largest development ever in New Caledonia's history, with plans to unearth about fifty-five million tons of nickel and cobalt mineral reserves over thirty years."

The Bechtel engineer came to life as he shared what he knew about development in the region, to offer some context to current mining practices on the island.

"Mining in New Caledonia began in the late nineteenth century, when an engineer discovered the first nickel ore," Andy launched into a brief history. "It has produced one of the wealthiest economies in the South Pacific, but definitely at a cost."

Mining in this Elysium environment for more than a century had led to considerable water pollution, soil erosion, and risk of endangerment to certain plant and animal species and coral reefs as a result of massive dumping of waste material down the slopes below the nickel mines, which then flowed downstream through rivers and tributaries, back to the ocean. Consequently there was ongoing resistance to large-scale mining operations by the island natives, whose ancestors dated back generations, long before Great Britain's famed Captain Cook landed on the island in 1775 and named it after his beloved Scottish Highlands. Environmentalist groups lobbied heavily and continuously for stricter policies to protect their wetlands and mangrove forests from further industrial degradation.

However, the Kanak people lacked legal sovereignty and therefore true political representation. Unlike their next-door Australian neighbors, the debate over indigenous self-governance on New Caledonia was still in its infancy. French authorities, for the most part, backed the Goro project and were avoiding constitutional parley with the Kanak. But it was only a matter of time before they could no longer do so.

"We're keeping an eye on the Rhéébù Nùù," Andy continued.

"Rhéébù who?" Terry interjected. Andy didn't skip a beat.

"A local environmental group that started up recently to monitor activity at Goro. Rhéébù Nùù means 'eye of the land' in the Djubea language."

Ted nodded, taking in the information, adding a new overlay to his understanding of the project.

"So far, they've been putting some pressure on us to consult with them, but they're not trying to shut us down. They don't have much political leverage at this point, so I don't think it's anything we can't handle," Andy concluded.

In his gut, Ted wasn't quite so sure. He thought of Voisey Bay and how the IBAs had paved the way for a positive working relationship with the Innu and Inuit. These days, you could never be too careful. If the sparks of dissent were there, it was only a matter of time before flames ignited.

Moments later, they were greeted by French officials. The Bureau de Recherches Géologiques had owned the original mineral reserves back in the '70s then sold them to Goro Nickel in the mid-1990s, but still retained 15 percent while the Inco affiliate held the lion's share.

As they toured and chatted about the project that day, two things became clear to Ted: If Goro was Shangri-La, then treated with extreme care, she might offer up her mineral riches. Handled poorly, her bounty could be locked away forever. As every good PM knew, no project comes with money-back guarantees.

But it was a team lead's responsibility to manage the issues and risks while keeping on target. This job was delicate. At the same time, it would require a rigor of dialogue, detail, and diplomacy that, so far, neither he nor Terry had witnessed at Goro's headquarters.

His boss was flying back tomorrow. Ted had agreed, with some reservation, to stay on for the week and see if he could advise the team and help put some order to the internal chaos. It was becoming quickly obvious to Ted that they had enough issues to manage on Goro without the added trouble of a warring team.

"Just 'til Friday. I'll give it my best shot," Ted had said. "But then I'm going home."

"Yeah, yeah. I know. You got it, Ted," Terry had responded, unconvincingly.

Truth was, Ted didn't believe he'd be let off the hook so easily.

He'd been brought in because no one else wanted to deal with the mess, and truth be told, Ted found it difficult to turn down a challenge. Even if it took him to the other end of the globe. The magnitude of this fast-derailing project was chipping away at his resolve not to become involved. He'd do what he could to get the project on track. Time and experience had taught Ted a few things. Maybe it was time to give back, offer some badly needed perspective and leadership. If they kept spinning their wheels this way, six months down the road, they'd be deeper in the mess.

But Ted had questions. He'd noticed that the government team had been quite cool and reserved during their site tour. Why was this? With the mine and its key partners all based in New Caledonia, why were Bal, Ron, and his disenchanted team all holed up in Brisbane, Australia? Wasn't that a recipe for miscommunication and disharmony? And as for Terry, the VP of projects, his sole purpose was to keep the projects in his portfolio on budget and schedule. This project was Down Under, indeed, Ted thought.

Whether Ted liked it or not, Goro had a grip on him.

***

Listening to Ron criticize the Bechtel Technip Hatch team was starting to fray Ted's nerves. He wasn't much for conducting closed-door meetings and less for stooping to play the "he said, she said" game. But his technique was working. By observing from the side lines, while creating a safe and confidential environment for each team member to offload, he was digging up dirt faster than a hydraulic Cat shovel working double shifts on a sleepless mine operation. Considering they hadn't broken ground yet, what were the options? There was no time to mess around. This was Mission: Damage Control. Finally they were getting somewhere.

Ted leaned in, across the table toward his worked-up associate, who was nursing a bout of verbal dysentery. Ted let him vent a moment longer and then finally cut in.

"Ron, who's in charge?"

"What?"

"Who's in charge?"

"Well...uh...you know, Ted, it's a good question. I mean, am I running the project or not?" He paused, hands gesturing wildly.

"Jeez." Ted fell back into his chair, lines emerging across his forehead. They sat a moment in silence.

On the plane ride over, Ted had scanned Ron's résumé. This was the biggest project of his career to date. He had not ingratiated himself to the engineering contractor.

"We've got Bal from Bechtel. And he thinks he's everyone's boss. The other fellas, from Technip and Hatch, they can't stop arguing long enough about their damn contract so that we can get 'er signed and get on with the program. Meetings have become a complete..."

"Ron." Ted interrupted him.

"waste of time."

"Ron!"

The two men stared at each other. Ted sat bolt upright, and Ron fell silent.

"Are you telling me that the deal between Inco and its principle contractors hasn't been signed? That you've been working here, under the same roof, for more than—what? Six months?—without a legally binding contract and more than a billion budget dollars at stake?"

Ron let out a long sigh and slumped in his seat, chin dropping to chest. "That's correct." His voice was strained and meek. "They just...won't...sign."

Ted couldn't believe his ears. An international consortium, Bechtel Technip Hatch (BTH) were a package deal hired on a "target price" basis to construct the mine. This meant that they would be paid the cost of work carried out—estimated at $1.3 billion—but would also share in any savings or cost overruns, less or greater than the tendered amount. In other words, they were sharing the risks as well as the rewards of this sizable job, so that was the reason they had their own PM on staff to copilot the project. Such tripartite ventures were one of the many ways in which big companies those days were trying to save on operation budgets, but in this case it was backfiring.

"These guys have their shorts in a knot over the definition of 'change.' BTH wants different language in the contract to ensure that they can make scope changes to the project without our approval," Ron explained. "Until they're satisfied, they've flat out refused to sign the contract."

The lawyers and managers had been unable to reach consensus for months.

"Well, it's not surprising they can't agree on it. Inco couldn't possibly live with such a clause in the contract," Ted said dryly. But if the big bosses knew the situation, they'd flip a lid. He touched his fingertips together lightly, forming a pyramid, as he tried to wrap his head around the implications of what he'd just heard.

Ron took this pause to ramble on about his concerns over the quality of people on board and the so-called dilution of his team by questionable subcontractors continually being hired by BTH without his approval. Bottom line, Ron said, he wasn't getting any respect around there. Furthermore, he was bogged down by scheduling changes, contract amendments, equipment holdups, budget revisions, cost overruns, and a niggling dispute that wouldn't go away with a local indigenous group over the environmental impact report included in their initial mine proposal. As Ron's mouth moved he glanced at his watch with a worried expression.

Ted had all but stopped listening. He was still stuck on "no contract." This was bad on so many levels.

As they spoke, money was being spent. Management had mobilized, and design work was underway, yet according to Ron's summary, the technology that they'd acquired to extract the nickel ore from a South African outfit was so new and cutting edge that there were still kinks to work out and more testing was needed before groundwork could even commence.

Meanwhile, supplies had shipped, and a workforce was in place, but construction was at a standstill, and community support sounded tenuous, at best, and at worst, nonexistent. Strained relations with local officials due to a decentralized head office were not helping matters. Ted knew projects hardly ever went according to plan, but this Down Under doozy was ass-backward. From the team who'd been assembled to the muddled scope and direction, Goro was a legal land mine in a contractual no-man's-land just waiting to explode. He had two choices: detonate it safely or get the hell out of the way.

"OK, Ron. Thank you. I appreciate your…candor. That's all for now."

Ron stood up and left. Ted stared at the phone, considering a moment whether to call Terry, and then he remembered it was 2:00 a.m. in Toronto. Let the fella sleep the night soundly. News like this could bring on insomnia, or worse. Besides, Ted figured he might as well get the whole story first. He had the rest of the team to interview. As he left the conference room and turned right down the hall toward the Bechtel project director's office, he glanced, momentarily, over his shoulder to see Ron slip into his own office and shut the door.

Unbelievable. Ted shook his head. It was like being caught in the middle of a bloody lovers' quarrel. What next? He braced himself. This could get messier yet.

***

Bal was swiveling in his chair, as Ted entered his office. Best not to pick sides, Ted thought, as he pulled up a seat. Bal flicked a ballpoint pen against his desk. He looked tired and strung out, ready for a fight. The walls oozed negativity.

Bal's boss strode into the room. Andy looked almost embarrassed as he set down a pile of folders containing specs and blueprints, and then glanced at Ted, raising his eyebrows. Bal shrugged.

"Thanks for joining us, Andy." Ted smiled and then got up and closed the door. Ted sat down and flipped open his Day-Timer. "Let's get to the point. Bal, in a hundred and fifty words or less, what is the problem?"

Bal instantly grew exasperated. "What isn't the problem, you mean. Ron's been operating his team in silos for months. Inco people don't come to our weekly meetings. When we had our all-hands AGM meeting, Ron didn't even bother showing up."

"Have you tried talking with him?" Ted suggested.

"Ted, we've tried talking to him on countless occasions—you know, to get some proper communication going and some alignment on things." Bal's voice had taken on a pleading tone.

Ted tapped his fingers. He wasn't sure anyone was playing on the same team. "Sounds to me like it's become every guy for himself around here."

Bal ignored Ted's comment and continued to vent. "We're duplicating work, there've been oversights and mistakes made. Nothing serious, but we can't continue like this. It's putting the project in jeopardy." With that, Bal sat back in his chair. He and Andy exchanged nervous glances.

Ted turned to his friend, who sighed.

"Andy, any thoughts? Do you agree with Bal's assessment?"

"Ted, I'll tell you a story," he began calmly.

"Good, let's hear it." Ted folded his arms across his chest.

"Terry came to me in Toronto, just after he'd married the BTH crew for construction work, and said 'I have five project manager candidates.' He gave me the list. He said, 'Andy, would you rank them for me?'"

"OK."

"I ranked Ron as fifth, and Terry went and hired Ron. Go figure."

Andy stretched out his legs and folded his hands behind his head. He looked like he'd had about enough also. Then Andy added a few thoughts. "Ted, the truth is, we're dying for direction from Inco. And we're stumbling along here." He leaned in closer to Ted.

Ted scribbled a moment in his Day-Timer. At least he was starting to get a clear picture of the problems they were dealing with. When he was done, Ted turned back to Andy, who'd started to look nervous.

"What's up, Andy?"

"Well, Ted, this opens up another can of worms entirely. I hope I'm not stepping on anyone's toes by saying..."

"Saying what?"

Andy shifted in his seat uncomfortably. Ted braced himself.

"I think the budget for this project was underestimated."

Ted put down his pen. "You've got to be kidding."

"I wish I was." Andy shook his head, and Bal just watched Ted as his boss continued with his explanation. "Now, we've got Ron on a cost-savings mission because his ass is on the line. Meanwhile, everyone's in-fighting over unresolved contract details. To top it off, no one even knows who's in charge—including the guy who's supposed to be in charge!"

"You think the project costs exceed $1.45 billion?"

Andy exhaled forcibly through his nose and his shoulders dropped. "Given the scope, technology, location, and politics of the project...unfortunately...yeah. I'd say, much more."

"Jesus."

For a moment, Ted was still. Then he slapped his hand down on the desk. "Well, if that's the case, then we need a new budget. If this is a budget issue of the magnitude Andy's suggesting, then you've all got to pull together, figure out a realistic budget, and ask for approval from head office. This is above management's scope."

Andy nodded, and Bal looked unsure.

"Look, fighting among yourselves and not looking at the big picture will just end up costing you the project and maybe your jobs. Let's turn this around, fellas." Ted spoke calmly.

Bal gave a half nod. "All right, Ted. That makes sense."

"Good."

Ted looked at his watch, and then he looked out the window, into the beautiful Brisbane afternoon. It was time to clear the air, help the team connect, and turn Goro around. Every project had its problems. Goro's were plentiful: low morale, poor communications, unclear authority, and an unrealistic budget, just to begin with.

Then there were other ticking time bombs: the politically oppressed Kanak people, who lived on the land near the mine site and had no IBA agreement with Inco, and the environmental group, Rhéébù Nùù, who were monitoring mining progress closely. It was only a matter of time before both exploded in their faces.

But for now, Ted needed to focus on saving the Inco-BTH management team. With any luck, a change of scene and a little group activity would help the team gel and bring some new perspective, so they could start working out together. Disasters were huge opportunities. At least Ted wasn't bored.

LESSONS LEARNED
1. Strong leaders result in good projects. Weak leaders result in bad projects.
2. The local community must be onside and management must be committed to the project location.
3. The Inco management team had limited experience and there were multiple reporting lines. This negatively impacted the decision-making capability.

CHAPTER SIXTEEN

Who's in Charge?

Ted landed back on Canadian soil late Sunday night. He barely had a chance to get settled at his desk with a cup of coffee Monday morning before Scott called and asked Ted to attend a meeting in the corporate office.

Inco's chairman and CEO asked, "How was Ozzie land?"

"You want the rundown?"

"Yep." Scott sipped his protein smoothie through a straw, and then set it down on Ted's desk.

"Well, Scott, to be frank, I did what I could in a few days but—the definition's not there. There's been a complete breakdown in communications between the Inco and BTH teams—if it was ever there in the first place." Ted paused for a breath. "And Inco's project director doesn't have any terms of reference. No one's told Ron what he's supposed to be doing. Is he running the project or not? He doesn't even know."

Scott didn't skip a beat. "Can you write up some guidelines for him—a list of ten things a project manager should be doing?"

Scott rubbed his jaw line, eyeing Ted intensely.

"Sure, I can do that right away," said Ted, reaching for a pen to make a note in his Day-Timer.

"Good. Let's meet first thing tomorrow morning." With that, Scott jumped up from the chair and made a beeline for the door. "By the way, welcome back," he said before leaving.

"Thanks, Scott." Back to basics, Ted thought.

Already generating bullet points in his head, Ted found a pad of paper, and then he didn't move for the next hour. When he was done, he folded his hands on his desk and scanned his list with satisfaction. Before him was a tight inventory of project leader guidelines for Ron. Part of due diligence, in Ted's view was being clear about the problems and addressing them promptly so the situation didn't deteriorate further. As he'd learned in Russia, and many times over, good people could save a bad system. But on the other hand, bad people couldn't make a good system work. People were people; they didn't change their basic stripes. The right team was so critical.

Then Ted set aside his list and forgot about it. Thankfully, Goro wasn't his problem anymore. For the rest of the afternoon, he reimmersed himself happily in the Voisey's Bay agreement. At four o'clock Ted grabbed his coat and left the office, stopping by the LCBO to grab some wine on his way home. Anita and his new son-in-law, Mike, were coming over for dinner. He was looking forward to putting his feet up and enjoying a couple of glasses of cab-sauv and some catch-up time with his family. The previous week's work and travel had been exhausting.

As Ted drove home, he did something quite unusual—for him, anyway. He let work drift from his mind, and in its place, Ted thought about absolutely nothing at all.

***

"No, this isn't any good." Inco's president heaved back in his chair, pushing Ted's list away from him.

"What's wrong with it? Ted asked, puzzled and confused.

Peter Jones turned to the CEO, Scott Hand, for backup. In that instant, Scott shrank in his chair.

"Ted, the first line: 'Inco project manager will direct BTH on this project.' First of all, he isn't going to direct them at all," Peter grumbled and then launched into an explanation. "We have BTH there on a target price. They're going to get a bonus incentive for doing good—coming in on schedule and cost—and a penalty for doing bad, and our team is just there to review what they're doing."

"Pardon?" Ted couldn't believe his ears. "And you think that's going to work?" He almost had to laugh at the absurdity of it. Suddenly he wished he'd recorded a video of the shenanigans he'd witnessed down in Brisbane so they could see it firsthand. "The team is stumbling over there. The BTH guys are looking to Inco for direction. And we've got to give it to them."

Ted pressed. There was something to be said for added efficiencies for cost savings, especially in a recessionary market, but their idea of tripartite babysitting was just asking for trouble.

Peter didn't like what he was hearing.

Scott nodded in understanding and then turned to Ted. "Which brings us to our next point."

He braced himself.

"We need you to go over there, and fix this."

Ted's hands flew up. "I just got back! What do you want me to do—move Down Under!?"

Peter jumped in assertively. "That's exactly what we want you to do."

Scott backed him up. "I'm not sure how else are we going to straighten this mess out, Ted."

Ted leaned in, folding his hands in front of him. "Listen, guys, when Terry hired me, I told him expressly that I'm done with international work. I've had it. I'm finished. Thirty years is enough."

"One more?" Scott petitioned.

They were asking for a lot. Such a cleanup mission wasn't a quick fix. He and Mona would have to move their life more than four thousand kilometers away. Just the thought of it exhausted him. Ted rubbed his forehead.

"Oh, Scott." Ted's voice fell flat.

Then Peter leaned in toward Ted and folded his hands. His voice was authoritative but calm. "Look, Ted, go home tonight and make a list of what you need for us to get you to go there. Truth is, we're desperate. We wouldn't ask if there was any other option."

Ted nodded, wondering what his wife would think. "I hear you. I'll talk to Mona and see what I can do."

***

Ted took the rest of the day off. He drove straight to the club to hit a bucket of balls. After a couple of minutes, he'd forgotten his day. The tension of the last week began to dissolve as he breathed in fresh air and swung his club. It felt good to move his body after so much sitting. Out on the course, it was just you and that little white ball. As long as you kept your eye on it, you had a chance. From there, your success depended on how well you read, and responded to, a series of calculations involving slope of terrain, distance to the green, wind velocity, and direction. Which club to choose? What speed and velocity of swing? Golf was a lot like projects. Both required precision, discernment, and your best efforts every time you stepped foot on a course. When the bucket was empty, Ted felt completely relaxed.

On his way home he picked up a bottle of his favorite 1995 Australian Houghton Show Reserve Shiraz. That night he and Mona discussed the proposal to move to Brisbane, Australia, where Inco's head office was located.

It was a complex wine—with fine oak tannins and a rich, full-bodied finish with strong berry flavors. But for Mona, it was a simple decision. She saw that her husband was torn, but, really, what did they have to lose? If he wanted to go, if he felt he had to go, then why not chalk it up to another adventure?

So Ted made his list. He had to have unique authority to direct the project; Mona could accompany him; all their accommodations and expenses must be paid; he'd have to come up with a number. Ted stopped short of throwing in a kookaburra. There'd be plenty of those in the Land of Oz.

***

The next day, Peter took one look at his list and stood up to shake his hand. "Done."

Scott simply said, "When can you leave?"

Ted and Mona were leaving for holidays in Phoenix later that week but could be ready to relocate by month's end.

"Good."

Then Ted brought up his boss. Terry had gotten him involved in Goro in the first place, but he'd been noticeably excluded from their discussions. He didn't want to step on any toes. "What about Terry?"

Peter spoke plainly. "Don't worry about Terry. He's VP of projects. He's got plenty of other fish to fry. You're on Goro now. But he's still your boss."

"And Ron?"

"That's up to you, but if you want Ron removed, we will organize it."

Ted nodded, satisfied. At least he would be in control to roll out the necessary changes before it was too late; otherwise, it was game over for Goro.

> **LESSONS LEARNED**
> A project needs clear lines of authority. Someone needs to be in charge and have the authority to make decisions.

CHAPTER SEVENTEEN

Diagnosis

"Dad, you can't be serious!"

Anita's face was flushed with indignation. For a split second, she looked like that child at the tender age of six who'd just learned from a boy in her first grade class that reindeer couldn't really fly.

"Anita, I don't have to go. We'll just stay here." Ted's voice was firm but undercut with softness.

His forehead rippled in waves of concern. Mona sat beside him, holding a tissue. Mike was in the kitchen fixing some tea. They'd flown in on a red-eye from Phoenix as soon as they got the news.

"It will be OK, Mom. We've got some of the best oncology doctors here in the country. We're going to treat this aggressively, and I'm going to beat it." Anita's voice was resolute, and it bolstered everyone.

She and Mike had tied the knot that summer. Following the oceanfront ceremony, the newlyweds had shaved their heads in honor of Mike's mom and a sister of one of their bridesmaid who had both died of cancer. Anita had come up with the idea to bike across Canada for their honeymoon, raising funds for cancer research along the way. While on that trip, she found a lump in her breast. A few months later, the doctors confirmed she had breast cancer in three lymph nodes.

Tears glistened in Mona's eyes. "You're a fighter, sweetheart." She smiled bravely.

Diagnosis

They all just had to pull together and be positive.

Anita put an arm around her dad. "I'll be OK. We'll be OK," she said, looking up as Mike walked into the room with the tea. "You guys need to go to Brisbane."

"Absolutely not."

Neither Ted nor Mona had slept a wink that night on the flight over. The news about their daughter's diagnosis made every other concern in their lives vanish. Nothing else mattered now but being there for her. There was no way Ted was about to relocate to the other side of the world.

"But Dad, what are you going to do here?"

Ted turned to his daughter. "What we're going to do is stay right here and help you with whatever you need."

Anita sighed. "Dad, I know how much you want to help. Thank you. But I know you. If you stay here, you're just going to mope around about me. Why don't you go and do your job? Take Mom. Show her the Outback. Keep living your lives, just like we intend to do. Mike will take care of me, and I will take care of me." Her eyes shone with determination.

Later that week Anita was scheduled for her first chemotherapy session. True to her warrior-like spirit, she had opted for the most aggressive treatment available to decrease any future chances of remission: a double mastectomy and surgery to remove her lymph nodes.

"She's got a point, Ted," Mona said gently, touching Ted's leg. "Maybe we should go for now, and just come home as often as we need to."

Ted slumped in his seat, elbows to knees, resting his head against his folded hands. For a moment, he sat there, unmoving. When he finally lifted his head, he looked extremely tired. He examined the front and back of his dominant hand and then twisted his wedding band and engineering ring as he mulled things over in his head. No job was more important than being there for his little girl. But Anita wanted them to go to Australia and keep living their lives. And if that's what she wanted, if it would strengthen her in the battle she was facing, then that's what they had to do.

Without a word, Ted turned to Mike, his eyes beseeching him to take care of his daughter. His son-in-law nodded. As if reading their thoughts, Anita reassured her dad.

"Mike will be with me through everything."

"Damn straight." Mike winked. "For better or worse."

Everyone laughed, and Mona wiped her eyes, and then gave Anita a hug. They'd be OK; Mike was solid, and he and Anita, together, could conquer just about anything—whether it was biking across Canada or day-by-day beating this cancer.

Ted put his arm around Mona. "I'll call Scott first thing tomorrow, and let him know the circumstances have changed...and that we need an open checkbook to come home whenever we need."

"OK." Mona smiled, her voice more buoyant. "I guess I'm going Down Under, after all!"

"Yay!" Anita clapped her hands at one small victory.

CHAPTER EIGHTEEN

Too Many Wrongs

"Terry, Goro has become a public relations nightmare."

"I know, I know. It's all over the papers here, too. That Rhéébù Nùù group is stirring up all kinds of problems. How's the schedule holding? Have there been any delays?"

The Goro project was nothing but one big delay. Ted braced himself. He was equally annoyed at having to brief his boss on a job he'd taken over the reins on six months ago. But as VP of projects, Terry's stamp of approval was required on any major budget or schedule changes. Terry wasn't going to like anything Ted had to tell him. So he eased in gently.

"Well, the team is on course, finally. I introduced them to streamlined and integrated data reporting so they can track all aspects of the project and monitor progress." Since Ted had stepped in, Ron had been removed from the project. The Inco-BTH team was finally on speaking terms—their roles having been clarified—and they were now working together effectively, managing the many ins and outs of a large-scale international mining project.

"Good, good," Terry replied.

Progress, however, was slight. If it was one step forward, one day, it was two steps back, the next. Meanwhile, big bucks were being spent.

Ted continued with his update. "But Terry, I'm afraid the backlash from not consulting in good faith with the Kanak people in the first place is getting ugly. They're digging through the Installation Classée and asking a lot of questions. It's stirring up negative sentiment, slowing down government approvals. I'd say, realistically, we're looking at several months before we can actually break some ground."

On the other end of the line, Terry said nothing more for a minute; he made himself busy shuffling papers around on his desk. The reality of the situation wasn't pretty at all. Just when the Goro project had started looking up, a ticking time bomb went off on this nickel-rich island oasis: the local activist group, Rhéébù Nùù, was up in arms, trying to block plans to dig in an ecologically sensitive area. The bad press was everywhere. The group was calling Inco's original environmental impact studies flawed. In recent months, they'd gone public, enlisting the support of several nongovernment organizations to put pressure on French authorities to stall the Goro project, pending an independent scientific review.

Ted was anxious to get the facts. Once he'd worked out the internal kinks with his team, he'd sat down with a copy of Inco's initial application for mining rights in New Caledonia and finally read it cover to cover. What he discovered was that the impact report was thick on safety and viability claims but thin on solid science to back them up. Furthermore, several key appendices were missing. His stomach had turned as he considered all of the potential reasons for this. Maybe they'd sidestepped due diligence and meaningful discussion with neighboring communities in hopes that pro-Goro factions would push the lucrative project through anyway. Or perhaps the science to determine the true impact of the Goro project just didn't exist yet. Any which way he viewed it, they were losing ground instead of breaking it.

Ted let out a heavy sigh. He'd been bound to his desk all week. On the one hand, he was grateful for the distractions of work as it kept him from thinking about his daughter's health struggle; thankfully her cancer was in remission, but they were taking it one day at a time. On the other hand, he was jonesing for a hard hat and a steel-toed tromp around an active mine site. But at this rate, they'd be lucky to move enough dirt to fill a sandbox by spring. For the time being, Goro's nickel would remain untouchable.

Ted pushed on with the unavoidable conversation. "Terry, I read the Installation Classée. I'm sorry to say, but it's pretty sad. Where are the supporting studies? And the IBA?"

"An impacts and benefits agreement? Ted, there isn't one."

"Why not?"

"It's not our fault. The French government isn't offering to share its piece of the pie with the Kanak people."

"Well, I think it's gonna backfire. Big time."

Terry cleared his throat.

Based on evidence, they had little fighting proof that Goro's megabillion-dollar nickel-mining operation wouldn't cause irreparable damage to a rare and protected ecological area of the South Pacific. Furthermore, aside from jobs, there would be minimal benefit to the local indigenous communities who'd lived off the lands for centuries.

Ted bit the bullet. "Now, about the budget—we need a new one."

There was a loud bang on Terry's end that sounded like a fist hitting the desk.

"Terry, I gotta tell ya—we cannot build this thing for $1 billion dollars. I've had a project controls guy I know take a look at where we are, realistically, in terms of cost and schedule, and he said, 'Ted, my best assessment is $2.3 billion, and you gotta add a year to the schedule.'"

"Well, I don't want you to publish that," Terry threw back, toxically.

Ted grew angry. He had no patience for wishful thinking or anyone unwilling to conduct business in an open and transparent manner; both killed projects and reputations. He'd rather walk than compromise himself. Ted's gut told him to hit hard or go home.

"Terry, you have a choice: we go ahead with the new estimate or pull me off the job." Ted's future hung in the pause.

"Yeah, OK. Go ahead. Do it."

**

The sun was shining through Ted's office window, but he was feeling cloudy. He hadn't slept well the previous night. Ted figured he probably had jet lag, having just returned from a minivacation in Cancun, Mexico, with his wife, Anita, Mike, and his son, Paul. But Ted didn't rule out the possibility that he'd been plagued by whatever had afflicted Ron in his final months on the project. Goro was taking its toll on him.

Ted sat hunched over his desk drinking a cup of coffee as he scanned the day's headlines. There were several mentions of Goro Nickel in articles detailing continued talks on the New Caledonian mine development project and mounting resistance from local activist groups and NGOs. None of it boded well for their scheduled spring launch.

Ted set down his coffee and looked out the window. He'd never felt like a project could swallow him up whole and spit him out. His team was nearing wit's end, as well. Now, to make matters worse, they had a new adversary on their hands. That morning, Ted had arrived at the office, only to learn that the entrance to their mine site had been blocked. But it wasn't the Rhéébù Nùù stirring up trouble this time. They were dealing with an entirely new—and possibly criminal—element. Word on the street was that the mafia had infiltrated the local stevedores, who ran the island port of Nouméa.

The only main point of entry by boat to the mine site, Nouméa and its docks was built by Inco the previous year as part of its initial contract with the French government. That morning Ted had received an emergency call from the Inco country manager, Pierre, who was shaking in his boots. An hour earlier, Pierre had been en route to meet up with the Bechtel engineering team when he encountered the human barricade: three large men with crossed arms and machine guns who'd shaken their heads menacingly at Pierre as he approached.

"Ted, these guys were armed, and they meant business."

"Christ. Is this a joke?!" Ted was incredulous. "First Ron, then the Rhéébù Nùù, now it's Godfather in South Pacific Paradise?" Momentarily, he thought someone must be pulling his leg.

"This is no joke, Ted." Pierre's voice fell flat.

Ted heard real fear in his colleague's voice, and then knew Pierre wasn't bluffing. Apparently, the bigger the project, the more that could go wrong.

"OK, Pierre," Ted said calmly. "Now, tell me exactly what happened."

That morning, around 7:00 a.m., the country manager had been attending to some business at the docks. Then about a half hour later, he'd made his way toward the mine site where he'd encountered three men just outside the entrance who'd refused to let him pass.

"The biggest guy, he was the leader, he told me the mine was closed today. And he said if I wanted it to reopen, I'd have to pay a visit to 'Roger' at the harbor Wednesday morning. I said, 'Who's Roger?' And one of his wingmen laughed, and he said, 'Roger's the boss around here.'" Pierre swallowed hard, as he relived that morning's events. "Ted, whoever these people are, they're obviously making a play to control the movement of goods on and off of the island. And if they really are the mafia, we seriously don't want to mess with them," Pierre stammered.

"Hold on, Pierre. Take a deep breath." Ted needed the facts, so they could figure out what to do next. Letting emotions get out of hand would not help the situation. "Now, start over and tell me exactly what they said to you."

Pierre exhaled heavily. "Well, the guy who was doing most of the talking, he said that this guy, 'Roger, has something he'd like to discuss with you, and your partners.' I asked him 'What about?' and he just said, 'You gotta go see him. He can make this little problem go away.'" Pierre paused, trying to regain his composure.

"Go on." Ted prompted.

"Then he said something about some kind of 'mutually beneficial arrangement between us and them.' He said, 'You'd better be there Wednesday.' That's about it. Ted, I didn't know what else to do. So I told him I'd show up at the dock."

"Jesus."

"What else was I supposed to do? They had machine guns!"

"Pierre, you did what you thought was right. But you know we can't get involved like that. We'll find a way, but it won't be their way. Whoever they are, we're not getting involved."

His colleague sighed again on the other end.

"So, where are we at, now?"

"Basically, nothing's getting in and nothing's going out. We're at a complete stand still. Ted, what do you think these guys are capable of doing?" On the verge of becoming unhinged, Pierre looked to Ted for answers.

"I don't know, Pierre. But I think you're right: they want to control us and everything coming on and leaving the island, most likely so they have free reign to smuggle contraband in and out of New Caledonia."

"What are we going to do?"

"Well, first of all, can you make your way to Brisbane immediately for an emergency meeting?"

"Sure, Ted, I'm on my way." Pierre sounded relieved.

***

That afternoon, Goro's team leads met to discuss the emergent problem. Ted was adamant in the meeting that they stand their ground.

"They're vying for power with us, but we're not giving in to these guys, whoever they are. If we do, the job's sunk."

Nobody wanted to cross "Roger" and his beefy henchmen, least of all Pierre, who'd felt their threats firsthand. He had a few things to say about it. "Look guys, I'm not going to lose my fingers, or maybe a few toes or a head over a few shipments of titanium. Why don't we just give them what they want and go about our business and let them do theirs? It has nothing to do with us."

"That's right, Pierre. It has nothing to do with us, and it has to remain that way." Ted put his foot down.

Over the last couple of hours, Pierre had grown increasingly agitated and troubled over what might happen if he stood Roger up at the docks. The other team leads looked skeptical and turned toward Ted for direction.

"We are not going to allow anyone to control our port, especially if it involves criminal activity, if that's what's going on here." Ted looked at each of his managers in the eyes, hoping to send his message home. "We built that port, dammit, and our names are on this project. We simply can't allow this to happen."

Pierre didn't look convinced. "Well, that's all good in theory, Ted, but..."

Ted shook his head firmly, and Pierre fell silent. "If we just lay down and do what they ask, our credibility is shot with everyone. We'll expose the incident for what it is, and with any luck, they'll back off."

If they refused to be bullied and sent a clear message to their antagonists, then maybe they'd have a leg to stand on and Goro could get back to business as usual. They took a vote and all agreed to stand firm and ignore the intimidation. Pierre reluctantly fell in line with his team and raised his hand.

In truth, Ted wasn't sure how big a threat they were facing. He made the best decision he could, at the time, and hoped for the best outcome. The situation required leadership, and so he stuck to the moral high ground. He had no prerequisites in his career for handling the mafia, though the Saudis and Kyrgyz had taught him to steer clear of anything that smelt of a bribe.

Unfortunately, or not, Pierre did what he thought best, anyway. By the time Ted heard what his colleague was up to, there was nothing he could do. The next morning, breaking team protocol, Pierre went down to the port and paid a visit to Roger. He cut a deal with him, and the next day, at the mine site, it was business as usual.

LESSONS LEARNED
1. Ensure the project is adequately defined prior to making large financial commitments.
2. Don't fall in love with the project. Inco management wanted the project so bad their decision-making ability on cost and timing was clouded.
3. A successful project requires an accurate cost estimate and a realistic schedule.
4. You must have environmental issues resolved before proceeding with the project.

CHAPTER NINETEEN

La-La-Land

"So, what happened?" Jim clutched his cold drink, leaning forward in his chair.

Ted wiped a bead of sweat from his temple. The heat was stifling, despite the shade of the veranda.

Enthralled by his cousin-in-law's story, Jim adjusted his baseball hat, waiting for Ted to continue.

"That day, Pierre put a Band-Aid on the situation, but the project was already hemorrhaging from too many other places," Ted explained.

"You mean the Rhéébù who?"

Ted laughed, and paused to take a sip of his white rum and Diet Coke. He was thoroughly enjoying himself. "You got it."

That spring, the Rhéébù Nùù organized a major protest in the capital city of Nouméa. More than three thousand indigenous and local dissenters of the Goro mine project attended the event. The French military broke up the protest, and there was some police brutality.

"It was really bad. People got hurt. The story was everywhere." Ted studied the condensation on his glass thoughtfully.

It was all downhill from there. The groups' coordinated efforts put pressure on the government, which led to an independent review of Inco's initial permit application and, along with it, their flimsy feasibility studies. Ted shook his head recalling the verdict.

"What they concluded was that the Goro project, as conceived, presented too many risks to the country's fragile ecology."

"I could see that coming."

"Yeah, no kidding. Me, too." Ted shrugged, and then a good-natured smile crossed his lips. "Ah, you win some, you lose some. Goro made me appreciate the good ones." Ted sat back, crossing his legs.

Inco had started off on the wrong foot. They'd hired the wrong people. They'd failed to build good-faith relationships with the indigenous communities. The science to back the project was weak; even the exciting, new technologies they'd proposed to extract the hard-to-get-at nickel ore had never been tested on a large-scale operation like Goro.

"On top of that, get this." Ted raised his voice dramatically.

Jim raised one eye.

"The new budget came in, not at two billion, but a whopping 45 percent higher than the original estimate."

"Holy mackerel!"

"And just like that," Ted brushed his palms together, "upper management stopped the job."

Jim almost barreled over in his seat. "After all the millions spent—they just killed the project?"

"Well...not exactly."

There was more to that story. Ted adjusted his glasses. No one, including the Rhéébù Nùù as it turned out, could let that nickel ore alone. It was worth too much. Everyone wanted a piece of it. After all, it was their fabled Shangri-La.

"What can I say, people fall in love, and you know what happens?"

"What happens?"

"Their brains fall out."

Jim chuckled, plumping a pillow behind his back.

It was true. Inco management had wanted the project so badly that their decision-making abilities on cost and timing were compromised. On top of that, they'd overcommitted financially before the project was adequately defined. Management was not committed to the location, and the local community was not on the side of the project. Furthermore, the Inco management team had limited experience, and early on, multiple reporting lines had confused matters and negatively impacted their decision-making capability.

"In the end, Terry and Inco parted company. The project got suspended, pending an internal review, and then was eventually restarted with a new project team. I was offered VP of projects and took the promotion but got transferred home soon after."

"Oh? Why was that?"

"We were ready. Scott wanted me back in Canada, anyway, and Anita, you know, had taken a turn for the worse. Mona and I wanted to be close."

"Right, of course." Jim lowered his eyes a moment, studying his hands. He looked up again at Ted. "But she bounced back after that, didn't she?"

Ted's eyes brightened. "She did!"

He'd been amazed by Anita's fortitude during her battle with cancer and how she'd pushed on and beat the odds, day after day. For a moment he said nothing, looking out at the freshly manicured backyard. The gardeners had come that morning.

Jim changed the subject. "So, in the end, did the mine get built?"

Ted cleared his throat. "Yep. Some four billion dollars and nine years later."

Jim whistled.

"But word is, they still haven't produced a single ounce of economic nickel oxide."

"You're kidding."

"Nope."

"Holy Toledo." Jim looked up at the ceiling, chewing over this new piece of information. Then he looked back at his cousin-in-law perplexed. "Even though the project was suspended?"

Ted explained. The French government had quietly pushed construction through while Inco worked on building its case for the mine. "They tried to keep it under the radar, but everyone close enough to the site knew what was going on."

To that day, the Kanak people were still fighting for self-governance, stiffer protection laws, and, now, for their share of mining royalties on their native lands. Support had shifted in their favor. In 2010, the Toronto Stock Exchange had removed Inco from its Good Practices Index for failing to meet their human rights criteria.

"You're lucky you got out when you did, my friend. Sounds like that project would've eaten you alive."

"It almost did." Ted sighed. "I lost sleep over Goro. It was the biggest disaster of all time!" Then he grinned widely. "But Mona and I sure drank some good wine and played on some pretty fine courses down there—though I can't say I improved my handicap much. That was some rough sailing on Goro."

Jim laughed, folding his arms behind his head.

Goro taught Ted that big projects were the same as small projects; except that when things went wrong, big projects came with even bigger, more costly, problems. Some you could solve, if you caught them early enough, or avoid altogether. Others were like cancers. They spread and took over.

Jim cut into Ted's thoughts. "The moral of the story?"

Ted laughed, and raised his glass as he spoke. "Don't live in la-la land."

Working in the Land of Oz on Goro confirmed for Ted that, indeed, there was no place like home; home being reality.

The two men clinked glasses.

Ted added an afterthought. "You have to have the right team from the start. And you have to examine the issues and manage them in a real and transparent way."

Jim finished his thought. "Otherwise they'll surely come back to haunt you."

Then Ted thought of Anita. One of the greatest lessons his daughter had taught him during her battle with cancer was that adversity is a teacher: it shows you who you are, what you believe in, and where you draw the line.

Ted's forehead creased, and he turned toward his dear, old relative. Jim just nodded and smiled.

Sometimes a Band-Aid and all the love in the world still aren't enough.

LESSONS LEARNED
1. Sometimes what looks like a good project isn't. Don't fool yourself.
2. It's important not to be emotionally involved with the project. You must have a clear head to make reality-based decisions.
3. Some projects have it all: defined scope, a realistic schedule and cost estimate, and a great team. Some projects just don't have it.

CHAPTER TWENTY

Courage

Most people couldn't tell that Anita was sick just by looking at her. She didn't smile; she radiated life. Doing something halfheartedly was not in her repertoire, including how she battled her cancer.

When she was diagnosed with ductal carcinoma in 2001, the cancer had already spread to her lymph nodes. In that first year, Anita had a double mastectomy, several rounds of chemotherapy, and radiation. To give herself the best chances, she took charge of every aspect of her health, from sleep to diet to exercise. She asked for the support of her closest friends and family. And life went on.

Anita continued her accounting work in Richmond, choosing after a while to go part-time to conserve her energy for her other extracurricular passions. Living life to the fullest was always something she'd embraced; now she was its poster girl. Anita joined the BC Cancer Foundation to help raise funds and awareness for cancer research and became a proud spokeswoman for the Annual Abreast in a Boat dragon boating regatta and an eager participant. She kept curling with her competitive ladies league. They had their sights set on the nationals and one day the Olympics. Anita found a hero in Olympian and cancer survivor Lance Armstrong, and two years after her diagnosis, she founded team LiveSTRONG for the Weekend to End Women's Cancers. She was relentless, when she could be.

Then her schedule would be wiped clear, her entire world quarantined and confined to bed rest, as she underwent a new round of chemo. During these times, her husband, Mike, was always close by to help her through the worst—make her laugh, keep her company, or bring her one of her favorite immune-boosting acai smoothies. Seeing them together like this, Ted had thought, more than once, this is what geriatric love ought to look like one day for him and Mona, not for two young sweethearts just building their lives together.

But a week or so later, when she got her doctor's OK, Anita would be back throwing rocks at the curling club or off to deliver a speech at some event or fundraiser. Her positive energy and focus were electric, and her love of people and belief in the cause galvanized others to join the fight. All of this gave her hope, helped her heal, and inspired people around her, including Ted.

At last, the tests came back clear. The cancer was gone. She was in remission. They were all so grateful. It was over, or so they thought.

***

True Blue. Ted's eyes traced the words airbrushed across the stern of a yacht parked at the West Beach marina. He wondered what would he call a boat if ever he were to christen his own sea-worthy vessel. *Dream Project* maybe? The corners of Ted's mouth turned up. He liked the ring of it. Or maybe something with a bit more grit, a bit less pie in the sky, like—*Tough Love*.

That was more like it. A ship by that name could sail through a typhoon or gale on a stubborn sense of accountability alone, without the slightest promise of a sunset. Besides, after all of his years working in management, Ted knew that dream projects didn't really exist. He'd been down that road before—on Goro, in Siberia, in Kyrgyzstan, and way up in the Tien Shan Mountains.

You could do your best, set a clear course, bring aboard the right crew, prepare contingency plans, bear down and switch gears if the project started to stray off course. But sometimes people just dropped the ball, helicopters crashed into mountains, politics or economics or protests got in the way. Sometimes you came in midproject, and there was no wiping the slate clean. As with life, you could forge on, clean up the messes, but you didn't get any redo's. There were no guaranteed happy endings.

Ted glanced back at the boat. Anita. He could name his ship *Anita*.

But no. Not today. He couldn't go down that road, name an imaginary boat after his living, breathing daughter. She'd scold him from the other side of the Pacific for being so melodramatic. As he sat there quietly, he could almost hear her voice.

"Dad! You're doing it again! A small animal could get lost in there!" Anita pointed at the tracks of worry lines on his forehead. Then without the hint of a smile, his daughter looked at him, from her hospital bed, and told him to have faith; that she believed, and she needed him to believe too. Right then, Ted told himself to focus, stay positive, keep believing, and continue on. What he needed, was the kind of Olympic courage he'd seen in his daughter's eyes when she'd dared to beat the odds.

Then Ted sighed, smiling sadly, as he looked out from dock's edge where he stood on the suburban seaside port of Adelaide. Back in Australia, again. For one final project.

***

One spring day in 2005, Anita woke up and felt pain deep in her bones. Emergency scans revealed that the cancer had spread to her ribs and spine. There was no denying the diagnosis.

Ted had never felt so helpless. Anita was crushed. That day, he watched his little girl fall apart but only briefly. Once the agony of disappointment had passed and the pain had been managed, he witnessed her extraordinary spirit take charge of the situation. The major setback had a redoubling affect. She cut her losses and changed her game plan. There were still options, possibilities. Hope.

For ten months, Anita did more chemo. She resigned from the accounting firm and instead worked occasional shifts as a food controller at Sun Rich Fresh Foods, where she was surrounded by the local, organic, whole, and non-GMO food that helped keep her strong for her battle. Through this period, Anita stayed positive and active. Things were looking up. Her blood counts had stabilized, and her body was warding off infection.

Then the chemo drugs just stopped working. At that point, with the help of her doctors, Anita reassessed the situation and decided she had nothing left to lose. She entered herself in a phase-one trial for a new drug called Arry-380, which was credited with longer survival times for women with terminal cancer. Back in 2001, when she was first diagnosed, that drug simply didn't exist yet; her own fundraising efforts and belief in research for a cure had helped make it possible for her oncologist, Stephen Chia, to test and bring the drug to trials in time for Anita. The drug was their last hope.

***

Ted took a deep drink of fresh sea air, and checked his watch: 5:25 p.m. Mona would be arriving any minute. Then instead of exhaling, Ted held his breath. In a way, he'd been there, in that same place, a dozen times before, conflicted over a project. How was Olympic Dam any different than the others? This was his last overseas project, for real, this time. Over several weeks and bottles of Australian wine, Ted and Mona had made the decision to go.

Ted's reputation as the straight-shooting "cost" guy on Goro had led to him landing the biggest uranium project in the world with the corporate goliath BHP Billiton, smack dab in the middle of a global recession, when everyone was feeling the crunch. Apparently, the project needed some tough love, and he was the guy to deliver it.

He was excited by the offer, which was the biggest of his career, but he wasn't going to take it. Then, once again, Anita had begged for him to go. She had her life. They had theirs. Each day was precious and meant to be lived. Unable to curl competitively any longer, she had taken up coaching, and when she wasn't resting, or writing a speech, or walking for a cure, she was busy preparing her ladies for the nationals.

"Dad, this is huge! You should go. You have to. Please. For me?" Her eyes had shone brightly. And then he gave in and won one more victory for life. She'd wrapped her arms around his neck and he'd swallowed hard. Her cancer was still there, but you couldn't see it.

Ted's throat started to ache. Unable to hold his breath any longer, he let out a giant sigh and his shoulders sank a few millimeters. His face was a maelstrom of mixed emotions. He strode a few more paces, watching the late-afternoon tide rolling in against a sandy beach and partially overcast sky. The salty breeze felt refreshing on his skin and the crunch of sand beneath his loafers soothed his nerves.

Suddenly, Ted heard his former boss, John Somers, in his head: *You might as well aim high 'cause you're going to hit low enough anyways.* No matter the situation, John had always been skillful at walking the line between idealism and realism. Through the years, its project peaks and pitfalls, Ted had tried to do the same.

A perfect financial storm was brewing. Commodity prices were slipping daily with no signs of recovery. Some analysts were calling this the Great Recession and predicting the worst economic downturn since the Great Depression of the 1930s. Mining outfits from Russia and Europe to America, Asia, and Australia were consolidating their holdings, buckling down, and counting their liquid assets. Now, more than ever, cash flow was king. The diversification power of megacorporations like BHP Billiton put them in a unique position to turn huge profits despite hard times.

Ted took a few steps toward the dock's edge. His forehead was corrugated with endless rows of speed bumps. Over the years, he'd dealt with many a starry-eyed top executive who was unwilling to deal in real dollars and cents. In this business, you had to spend money to make money. BHP Billiton had the money but was unwilling to spend it. Tough times called for tough love. And it was Ted's job to deliver.

Looking out beyond the water at the soft pink trails low on the horizon, he wondered, after all of these years, if he had what it took to beat the odds on Olympic Dam. Money was not the problem. On any given day, the company was signing off on multiple budgets for mineral discovery, exploration of new reserves, project planning, R&D, construction, and labor and operations; servicing debts; and writing checks to governments and other royalty shareholders—all of which was strategically timed, budgeted, and prioritized. Meanwhile, investment capital was being pumped back through the fiscal turbines as revenue came in from the company's producing mines around the world; quickly earmarked for new or existing projects.

The philosophy at most companies, including BHP-B, was to minimize debt by paying for new capital projects with cash flow. Most outfits could not afford to do this entirely. Overextending themselves, especially during a recession, could throw a wrench in a smooth-running corporate engine. But BHP-B could do it. That last quarter, BHP Billiton was valued at a whopping $70 billion on the London Stock Exchange—more than double its perceived worth back in 2001, just before the strategic megamerger of Australia's BHP and South Africa's Billiton, which launched the company into a new corporate stratosphere. The Olympic Dam Project would use only one year's net profit to build. Yet with commodity prices falling, the head office was playing it safe and looking for ways to minimize cash outflow. In a nutshell: good project, bad timing. As the new cost guy on Olympic Dam, Ted's official job was to reel in the budget and steer the project and team toward success. Problem was, given what he knew and in light of the facts, he had no idea how he was going to do that.

Ted stepped back from the edge of the dock, and walked a few paces toward the marina. Then, as if unsure of his bearings, or even why he was traveling on that trajectory, he stopped, looked around, and then stooped his head, idly scuffing his topsiders along the wharf. His eyes fell to examining an interestingly shaped burl in the wooden pier. Then Ted's shoulders crept toward his ears as he recalled the terse conversation he'd had with his new boss, shortly after he'd arrived on the job. Ted had no choice but to tell him that the costs on Olympic Dam had been grossly underestimated. The news hadn't gone over well.

That day, Ted dealt him a reality check, while the Base Metals president was in the midst of his fanciful projections summary.

"It's a five-billion-dollar project, and it will be in production in 2013, and everything will be built..."

"Ah, it's not a five-billion-dollar project, Roger."

His new boss took the news like saltwater in an open wound.

"I'm putting this in the monthly report, you should know, Roger. I'm saying that it's a fifteen- to twenty-five-billion-dollar project. You'd better tell people."

"You can't do that!"

"Roger, I did it."

He kicked and sputtered. But finally gave in. Ted didn't like to be the bearer of bad news, but good companies paid a premium for tough love—even if they found themselves waist deep in water. But then rather than look for the proverbial hole in the boat, like any good engineer might, everyone started madly bailing. And Ted's boss passed him a pail.

***

It was 5:32 p.m. Ted scanned the marina. No Mona yet. A wave of fatigue washed over him, and he made his way to an iron bench a few feet away. Before sitting down, he pulled a folded, worn piece of newsprint from his pocket. Ted carefully opened an article from the *Vancouver Sun*, dated September 5, 2007, just six months back. His lips moved slightly as he reread the headline. The words were like warm milk to his frayed nerves: "New Drugs Keep Cancer Patient Looking to Future: Evidence of longer lives reported by BC Cancer Agency."

A BC Cancer Agency study suggests new drugs are helping women with an advanced form of breast cancer survive longer and live an average of 24 months after diagnosis.

Every day is precious for Anita Cochrane, a 36-year-old with a passion for curling, hiking, cycling, kayaking and dragon boat racing.

"I live every day to the fullest, even on a rainy day," Cochrane said in an interview. "Bad hair days have a whole new meaning. I love the hair I have right now because I've been bald so often."

Hunched in his seat, Ted looked up, smiling sadly. A picture floated into his head of Anita with a flower garland sitting atop her shaved head, the day she and Mike tied the knot, just weeks before she'd found a lump on her breast and their cross-Canada bike tour to raise funds for cancer research took on a whole new significance. What were the odds? Ted shook the thought from his head. He knew it was futile, that such an answer would forever be elusive. Dropping his head, he read on:

"I understand my prognosis and what I've been told, but there's definitely a part of me that knows I can beat the odds," Cochrane said. "Anything can happen. None of us really know what tomorrow will bring. And maybe tomorrow they'll find the answer."

The study, published in the scientific journal, *Cancer*, was coauthored by Anita's oncologist, Stephen Chia. Its results were encouraging. In the past, women with that advanced form of breast cancer survived about eighteen months. But the introduction of aromatase inhibitor drugs in the late '90s helped women live longer, by lowering the amount of estrogen available for tumor growth. Over that decade, the survival time rose more than 30 percent, from 438 days to 667 days. Anita's oncologist had called it "the first to demonstrate a significant improvement in survival time." Ted read to the end of the article, taking in each word like the Holy Scripture.

Cochrane said the study makes her more optimistic.

"The trend is in the right direction," she said. "Lives are lengthening. This decade will live longer than 24 months and, hopefully, the next decade will live even longer and they'll find an answer and we won't talk about death, [but instead] a chronic disease you manage and live with."

Cochrane added: "Every moment is just a real gift."

His daughter had two years, give or take, left to live. Dry eyed, Ted refolded the page and popped it back into his pocket, patting it for good measure. Then he stood up, brushed off his pants, and crossed his arms with new determination.

Olympic Dam was in trouble. And everyone knew it. The project had huge potential, but the timing couldn't be worse. No project was perfect. But few were beyond help. Ted would do his damnedest to make this one count. If there was a cure for OD's problems, he and his team had to find it—and fast. They had three days. Every minute counted.

LESSONS LEARNED
1. Always recruit the best people possible.
2. If there is bad news don't procrastinate. Deal with it immediately, or it will only get worse.
3. Sometimes what looks like a good project isn't, because the project doesn't line up with corporate financial objectives.
4. If a new project director comes on board when a project is already in process, it's important that an initial quick assessment is done to determine if there are fatal flaws to the project.

CHAPTER TWENTY-ONE

The Challenge

"So, the good news, folks, is that the mine is much, much bigger than originally thought."

Lead with the good news. Everyone knew, more or less, why they were there. The unpleasant truth, however, was in the details. For a moment, Ted just stood back, hands clasped behind his back. He was watching his expanded senior management team attack the impressive spread of gourmet luncheon sandwiches and homemade doughnuts, brought in moments before from the artisan bakery down the street.

"How big are we talking?" Brink inquired between mouthfuls of turkey-brie croissant, cupping a napkin under his lunch.

Through floor-to-ceiling windows, the Australian summer sun poured into their ultramodern meeting room. Ted pressed a button to adjust the blinds and lower the light. He and the bright, amiable South African engineer had become fast friends since Ted's arrival in Adelaide.

"Seventy-seven percent bigger. That's how."

Brink almost choked on his lunch. Others stopped chewing in unison.

"Holy Crikey!"

Now he had their attention. All heads turned toward Ted, eyebrows raised and a few jaws dropped open in astonishment. If it weren't for the bad news, he

The Challenge

might be enjoying this a little more. His team was primed, fed, and ready for the lowdown. Ted pushed his glasses up on his nose, cleared his throat, and dug in.

"When BHP acquired OD in the nine-billion-dollar takeover of Western Mining Corp. two years ago, everyone knew the mine was the world's largest uranium resource."

It was a bold move. Times were changing. The new Kyoto Protocol and Europe's carbon cap-and-trade scheme were driving the global growth of nuclear power—a relatively clean, reliable, and carbon-free alternative energy source. China and India were expected to lead the way, and the strategic acquisition by BHP would make the company a key player in the future of nuclear fuel. Ted flipped a switch and a bright blue line graph popped up on the wall to illustrate his next point.

"BHP has spent one hundred fifty million dollars since 2005 on its initial drilling program to prove up the ore at Olympic Dam—a drop in the bucket, really." The drill data itself was invaluable. "We now know that there are four billion tons of viable reserve in the ground—which makes Olympic Dam the second-largest mineral deposit"—Ted paused to add a level of suspense—"in the world."

Ted hit a button and the image faded. He turned back to his team. "This project is huge. Huge." He leaned in, pressing his hands against the table. "And the timing couldn't be worse."

Ted looked at his crew members, who were waiting for the other shoe to drop. For the next ten minutes, he painted a detailed picture of the company's corporate objectives on Olympic Dam: its short and long-term economics, the known constraints, and his senior management team's key role over the next three days in reworking the project's scope and direction to fit the new budget requirements. He'd need everyone on board. Each manager would have a hand in re-creating their project charter, and since they'd be looking at making some fairly drastic changes, they each had to sign it in order to stay on the project. This best practice helped keep everyone on the same page and working together toward a common goal.

Ted paused, took a sip of his lemon water, and looked around the table. He made clear eye contact with each of his five deputy project directors. Including their team leads, twenty-five senior managers were present in the room. Together they had to work a small miracle. By now, everyone was restlessly waiting to hear what kind of budget constraints they were dealing with. Ted didn't delay any longer.

"That's the good news, folks. OD has the potential for a sixfold increase in production."

"And the bad news, you say?" His quality-control guy, Jake Manson, cut in.

Ted nodded. Then he began, in a steady, assertive voice. "As you know, with the global financial market on its way down the tubes, we are here on a cost-savings mission." Everyone waited for Ted's next words. "An executive decision's been made to go back and study Olympic Dam, and it's our job to come up with a way to expand the mine, while keeping costs at ten billion dollars if not five."

Jake sat back in his seat, rubbing his chin. "Blimey."

Ted stood with his hands on his hips, watching the news sink in with his colleagues: a few deer-in-headlights, some nervous shifting, dubious looks, one high-pitched whistle, and a half-dozen blank faces.

Jake was the first to speak up. "Hold on, Ted. Didn't you just conservatively reestimate Olympic Dam as a fifteen- to twenty-five-billion-dollar project?" His voice was incredulous.

"Yes, I did."

"OK." Jake momentarily looked confused, and then he looked up at Ted's wipe board where he'd written a quote by Einstein: "Only those who attempt the absurd can achieve the impossible." A smile emerged on one corner of his lips as it dawned on him that his boss was in the same boat and just trying to make the best of the situation.

Unflinching, Ted looked back at his team. He knew what he was asking bordered on the absurd. They knew that he knew. Together, maybe with some luck and a small stroke of genius, they could make the impossible possible. That's what they all got paid the big bucks to do.

Ted strode over to the board, wrote "ideas" in big letters, and underlined it with one broad stroke. A jolt of excitement ripped through Ted's body. "We've got three days. Let's do this people."

Ted knew he had the best possible team at his disposal to tackle the problems on OD. "Let's give it our best shot. We've got all the expertise we need right here in this room—and tons of food—and after we're done, tonight we have a helluva fine evening planned down at the wharf for you and yours to unwind. So let's put our heads together and get things rolling."

The pep talk worked. Like any top-notch crew, everyone started talking at once. Ted took a deep breath and clapped his hands together solidly.

"And remember, folks, this is not a democracy, but everyone has a vote." A few guys chuckled, and everyone geared up for work. Then Ted adjusted his glasses, exhaled silently, and jumped in to chair the discussion. Fake it until you make it. In truth, Ted had no idea how they'd shave ten to fifteen billion dollars of savings out of such a beautifully distended project. Judging by the expressions in the room, neither did anyone else. At least they had a common place to start.

"OK. Who wants the floor?"

Jake jumped up from his seat. A tall, feisty guy in his early forties, the geological engineer had a longish face and sizable ears that stuck out from his short, reddish-brown hair.

"Great. What have you got for us?"

Ted sat back in his seat, reaching for a twist of sugary dough. He took a bite then washed it down with a sip of dark-roasted coffee.

By now, Ted was thoroughly enjoying himself. He'd asked Jake earlier that week to prepare a talk about past projects and how they'd tracked—what they'd done right and what they'd done wrong—so they could apply any new insights to their current project.

Jake began with a brief recap of other BHP open-mine projects then moved into an analysis of their current project. A science whiz with more than a decade of specialist experience working in BHP's geological departments, Jake was strong in the technical aspects but fairly light on management experience. Roger had recruited Jake as one of the senior managers well before Ted's arrival.

Over the next twenty minutes, Ted discovered his team's weak link. Whether it was nerves or just a dirt-poor attitude, Ted wasn't sure. As he listened, growing increasingly agitated, Jake told the group why he thought every BHP project had been a failure. Ted had no place for negativity in his boardroom. Giving Jake the benefit of the doubt, Ted waited for him to turn it around.

His half-eaten doughnut sat contorted on his plate. His deputy was on a roll.

"As we know, the existing smelter is fed by an existing underground mine. We are dealing with the most complex and inconsistent ore body in the world containing copper, uranium, silver, gold, and platinum ore. Now, here's where we're really stuck between a rock and a hard place."

Ted lowered his gaze and gripped his chair.

"We now have to move a billion and a half tons of rock, and, as we know, the only way to do so economically is to build an open-pit mine and expand the processing plant to handle it—in other words, spend more money to move more product in a depressed market with tighter purse strings."

Jake went on to tell the team that they wouldn't perform any better than past projects, especially if they failed to meet their cost and schedule objectives. At that point, Brink, the project financial guru took the floor. After explaining the overall project economics he closed by saying: "That's what we're dealing with, guys. So when this project gets stuffed up."

And that's when Ted blew a gasket. His hand came down loudly on the table. Then he jumped up from his seat. "That's quite enough."

Both Jake and Brink froze in their tracks.

"You know, Jake, I really wanted to hear about the good ones too."

A few of the guys saw it coming. The others looked on dumbfounded, stunned into suspended animation. Then suddenly they got busy reloading their plates, stuffing their faces with jelly-filled donuts, adjusting their notebooks, and clothing. Jake skulked back to the table.

A switch flipped in Ted's brain. Such negativity was a cancer to projects and teams. Not only did it destroy morale, it killed creativity. And Ted couldn't allow it. Right then and there, Ted decided to make an example of Jake. Beginning to pace, Ted's forehead broke into a musical staff of strained notes. When he stopped, he spun around to face Jake and Brink.

"Jesus Christ! What's the matter with you?!"

Jake and Brink shrank in their seats.

"We're not going to stuff anything up! We're going to succeed on Olympic Dam! What we don't need is an attitude of failure." Ted's piercing stare looked like it could burn a hole through Jake's head.

The rest of the team was shocked. No one had ever seen Ted Bassett spitting the dummy; he lost his shit, like a baby who'd just dropped his plastic nipple on the floor out of reach. Which made his outburst all the more effective. Everyone hung on his next words.

"Things go wrong, people. Projects go longer and cost more. Definition and scope change! But that's our business! We deal with the adversity and the change. There's no perfect project."

Ted spun around eyes wild. His body coursed with adrenaline. His voice was heard down the hall, full of frustration and contempt for complacency mixed with an unnamed anguish. This was on him. He had to turn this ship around. But there was no going back. Only tough love could move them in the direction they needed to go.

"We're not going to screw things up. We're going to handle this. And we will not accept mediocrity or failure. So stop telling me now how we're going to make a mess of it!"

By that point, Jake had disappeared into his chair. He was nervously running his finger and thumb repeatedly over his mustache.

Ted looked around the room then took a deep breath, adjusted his shirt and softened his voice. "Now, shall we?"

The tide instantly turned. Ted rolled with it, grabbing his wipe-board marker and heading toward the easel. He felt like he was floating, so he forced himself to concentrate on his feet touching the floor. He was conscious of everyone's stares. Momentarily he'd lost control, but apparently he hadn't lost any ground.

For the remainder of the afternoon, the team worked together constructively, and by three o'clock, they had several interesting cost-savings options on the table. One of them was to build a smelter in China and ship the ore overseas for processing. The idea was brilliant. And it had never been attempted before. Ted was riveted by it. Such a move could single-handedly change the scope, direction, and budget constraints on Olympic Dam entirely. Maybe even save the project.

He congratulated his team on their hard work and wrapped up the meeting on an upbeat note. "Great session, guys. Let's keep it up. And don't forget, cocktails at six thirty and a barbie to follow down at the wharf. Bring your significant others. And let's rage a bit, mates, then we'll get back at her tomorrow."

Grabbing remnants of their lunch, his team filed out. As Jake passed Ted on his way out, he nodded, averting his eyes quickly. The damage was done. Ted didn't feel good about it. But his strategy had worked. He'd cut off the toxic spread of negativity to his team. Many a friendship is made or broken in the boardroom. That day Ted and Jake's relationship pretty much ended.

***

Ted was jolted from his thoughts by Mona's voice. He turned to see his lovely wife approaching him from the other end of the dock. A blue shawl was draped around her bronzed shoulders. She had a bounce in her step. Ted felt a wave of relief as he watched her approach. She looked happy, having spent the morning on the golf course with a few ladies and the afternoon lunching at the club. He was glad she was adjusting to their new landscape. As Mona arrived by her husband's side, they linked arms and Ted planted a kiss on her head.

"How was your game? Did you break par?"

"No. But it sure was a beautiful course." His wife beamed. "I shot 96."

"Not bad. And the other ladies?"

"Good, good. We chitchatted the whole way around. It was fun. I found out Janice is doing her master's in fine arts and Krysta runs a book club. She only lives a few blocks away from us. I think I'll join. Right now, they're reading *Life of Pi* by Yann Martel. And how was your day, Ted? You look stressed out!"

"Yeah. Nothing that a cold pint and beach barbie can't fix."

Ted looked out at the water, where a gaggle of novice surfers chased the wake of crashing waves, riding, falling, and getting back on their boards. Mona studied her husband's face a moment quietly then followed his gaze toward the surfers.

"They look like they're having a blast out there."

"This is a good beach for beginners 'cause the waves aren't as strong as they are further down the coast. Maybe we should rent a couple of boogie boards one weekend." He winked at her.

"Well, I don't know about that, but we should take a drive down to Victor Harbor one of these weekends and watch the pros. Paul might enjoy that more when he comes to visit. Oh, and I talked to Anita an hour or so ago."

Ted looked at his wife with full attention.

"She's doing well. She and Mike were out on their bikes this morning, and she said she feels fully recovered from her last round of chemo." Mona's voice was triumphant. Ted's mood instantly lifted.

"That's really good to hear." He nodded, eager for more information, and Mona continued her update.

"She's got curling tonight, and she's practicing her speech for the Weekend to End Breast Cancer. She read it to me. It's excellent." Beaming with pride, Mona shook her head. "I honestly don't know where she gets her energy from, Ted. That girl has so much fight in her."

For just a moment, Mona's eyes began to cloud over, but before Ted could squeeze her hand, she recovered. Her face softened into a smile, and she asked Ted to point out which direction they were headed for dinner. As they strolled along the boardwalk, side by side, Ted told Mona about Day One of their meeting.

"I spit the dummy." Ted lowered his head in mock guilt.

"You? Spit the dummy?! Like a little baby??"

"I did. I spit it. At Jake. He deserved it."

"Oh, no. Poor Jake." Mona laughed. "What did he do? It must have been bad."

"It got everyone's attention; that's for sure."

Ted reached for his wife's hand, and he told her the whole story as they made their way to dinner. Now that the adrenaline had worn off, he felt a little ill at ease over the whole situation. Olympic Dam was pushing his buttons, taunting him like an immutable ocean wave in a raging storm. It was business, so why did it feel so personal?

Ted glanced sideways at his Mona as they talked and walked. Like so many times before, he admired her upbeat attitude, despite all they'd been through over the years. Now, more than ever, she was his anchor. Where had their beloved daughter gotten her spirit and courage? The answer was right there in front of Ted. Tenderness welled up in Ted's eyes. A warm wind blew at their backs, moving them gently forward. Each step they took was a lesson in letting go.

> **LESSONS LEARNED**
> 1. The project manager must demonstrate leadership and provide team motivation.
> 2. Make sure you know the project constraints before you set the scope. Otherwise, you will have to start over.

CHAPTER TWENTY-TWO

Ideas

Ted woke up at the crack of dawn. He felt more refreshed than he had in months. Jumping out of bed, he jogged along the beach for one hour and then had a shower and shaved. As he stood in front of his reflection, looking for any rogue whiskers, he paused and stared directly into his own eyes. For a split second, he heard John Somers's voice: If you wish to see the type of organization you manage, look in a mirror.

His old boss had spoken those words, more than a decade ago, during an off-site team-strategy meeting similar to the one Ted was leading now. It was 1997. In the wake of Bre-X, every corporation was madly churning out "best practices" policy and guidelines. Talk of ethics was on every off-site and AGM agenda. As new industry watchdogs, the Toronto Stock Exchange (TSE) and Ontario Securities Commission (OSC) were busy changing the rulebook on how new mineral reserves got reported to help prevent another Bre-X from ever happening again.

As Kilborn's head guy, what mattered more to John Somers than stricter policy and policing was how each and every member of his executive team conducted themselves when no one was watching. That, to him, was the difference between run-of-the-mill honesty and true integrity. John was a smart man, but more importantly, he was fair and principled. Furthermore, he believed in the people around him and knew how to bring out their best. Like a corporate samurai, he could cut swiftly through to the root of any problem. In a way, he'd been with Ted on every job site and in every conference room since they'd worked together at Kilborn. Ted wanted to be that kind of leader. So whenever he heard his mentor's voice in his head, he didn't question his own sanity but paid close attention.

Ted looked in the glass and listened. With bad news, procrastinating was the worst thing he could do. Olympic Dam needed swift and decisive action, or the situation would only get worse. Or else the true costs of this untimely megaexpansion project would get quickly out of hand. Right now, where things stood, Olympic Dam was a fifteen- to twenty-billion-dollar abyss in the making. If the goal was to cut costs, then the first thing they had to do was get out of the hole. They had to stop digging.

The first thing Ted did the morning of day two was stand up in front of his team and apologize to Brink and Jake for spitting the dummy. Teams expect their project managers to be dictators. They don't expect them to say "sorry." From there, everyone relaxed and opened up, including Jake, to some extent.

"So we're not going to dig a hole to China, folks, but why not build our smelter there?"

Day two. They were back at it, after a fun night of socializing and liquor-fueled antics at the beach. That morning, ribs and one-liners were still flying over their boardroom breakfast of fresh-roasted coffee and gourmet fruit-filled scones and turnovers. Marco Herrera had taken the Olympic Gold medal in the limbo contest, and their mining deputy, Jon Gilligan, had given a memorable impromptu toast full of witty and cheeky Down-Under puns. Even Jake looked fresh-faced.

In one hand, Ted held a dry-erase marker, ready to make some magic.

"Brink! It's a compelling idea. Tell us more."

A civil engineer educated at the University of Michigan, Brink Van Schalkwyk was a smart and business-savvy guy who'd worked as a fiscal auditor on BHP projects for seven years. He grasped both the technical and cost aspects of a project but also had a flair for original thinking.

"Here's the thought, guys." Brink cupped his hands together and breezily laid out his idea. "We could ship our raw material by rail, load it on tankers, and send it east for processing at our very own overseas plant. A smelter in China."

He scratched his head, pausing a moment to collect his thoughts, and then continued. "It would be a fraction of the cost—cheaper labor, and of course, we'd still benefit from the same commodity prices. Coca-Cola, Motorola, Volkswagen, Nike—they've all been manufacturing in China for years. But it's never been done before in our industry. With profit margins shrinking at home and the recession hot on our tail, maybe now's the time, yeah?"

Eyebrow raised, Brink glanced around the room. Ted had to admit it was a sweet idea. Given their cost-cutting mandate, the economic climate and current price tag on OD, it could really throw them a lifeline.

"It's a whale of an idea, mate," the deputy mining director jumped in. "If the goal's to maximize productivity and minimize cost, then a China smelter is practically a no-brainer."

"Yeah, Doc, then why didn't you think of it?" Marco Herrera, the project controls guy, aka the "Lad of Limbo," jumped in to razz his colleague.

Throwing his head back, Jon Gilligan laughed good-naturedly at his colleague. The senior geologist had a PhD and twenty-five years international experience in mining and exploration. He was as comfortable in a suit and tie as with thongs on his feet and a hammer in hand. Jon ran a hand through his hair, eager to get in on the discussion, and his tone shifted from debonair to deeply focused. Above his long, sharp nose, his brow creased in concentration.

"China's already the world's biggest user of metals. In 2005, they made up over 10 percent of BHP's total revenues, just from exports. I reckon they're not gonna slow down any time soon." Jon pressed his long fingers onto the table to stress his points. "To be there already sure would help us gain a stronger foothold in China's copper market."

There were nods of agreement around the room. China's rapid economic growth and the explosion of its middle class now accounted for one-fifth of the world's copper demand. The metal was used in everything from power cables and water pipes to car radiators and jewelry. Olympic Dam was loaded with copper. Since the late '90s, the mine's existing operations had produced more than two hundred thousand tons of it per year for global export.

This was the kind of exciting discussion Ted was hoping for. His team was on fire now. Ted sat back and listened.

Jumping in with measured optimism, Marco added his two cents worth. As the project controls guy, he was known for bringing balance to discussions. "With demand expected to outstrip supply in the next five years, BHP would be in perfect position to hand-deliver the raw goods," he began, enthusiastically. "It's a fine idea, clearly with many benefits, but I think we should be careful to observe due diligence when it comes to China's mining health and safety record. It's one of the worst in the world."

Marco went on to note that China had upward of ten thousand mining-related deaths a year. "I'm just saying. What you don't know can hurt you. Overall,

I like the idea. It seems economically sound. But that's pretty abysmal, ya know?"

Marco summed up his position. "We'd definitely need to ensure strong controls and monitoring in these areas. As long as we can manage the quality, this could very possibly work."

For a moment, everyone mulled things silently. Ted's dry eraser squeaked against the board as he added "access to copper" in bold type. Marco reached for a cherry turnover and leaned back in his chair.

Then Ken Sutliff, the ore-processing manager, piped in. "As they say, if you fail to plan, you are planning to fail. A little risk management saves a lot of fan cleaning." The veteran of twenty-five years was always ready with a quick joke or project management truism.

"Been there, done that," Marco replied, licking his lips and grinning. "And fortunately I survived it."

A few sniggers. Ted turned back toward his team enjoying the banter. He typically had a quip or two handy himself, in case a meeting became too charged or he wanted to lighten the mood or connect over something other than schedules or budget items. Spending hour after hour in closed quarters with such a diverse mix of people, personalities, and professional backgrounds left plenty of room for clashes and misunderstandings. A little PM humor, thoughtfully sprinkled, could shed some light or diffuse a situation, while reminding the group what they have in common. Ted tossed one in, in agreement.

"If you don't attack the risks, the risks will attack you."

Then Jake let one fly, hitting the nail on the head for their situation on OD. "Nothing is impossible—for the person who doesn't have to do it."

"Touché, Jakey," Marco replied, clapping his hands. And they all joined in.

Sometimes it felt a little hokey or premeditated, but there was no arguing that it worked. Whether he told a groaner, an eye-roller, or a real knee-slapper, at least for a moment, it wasn't a budget chart, scope debate, or schedule item. Laughter was a universal language, and it let everyone take a breath and press the "pause" button. Then it was back to business. Ted took the lead.

"So, let's come up with a plan, and do our best to keep it on track. Remember there are pros and cons to any idea."

He drew two columns on the wipe board, inhaling the ink with secret delight. They wouldn't even be there, exploring the options on OD, if the true costs of the project had come out earlier. It would have been shelved or shut down from the get-go. These three days were a lifeline. He turned around to face his team, his eyes more serious.

"At least we have the facts, and now we can see where this thing can go."

Everyone rolled up their sleeves. For the next hour, they worked through a list of pros and cons and a quick-and-dirty cost-benefit analysis. The savings from building an ore-processing smelter in China would be exponential. In fact, they could probably get their costs down close to the $10 billion mark and bring it within budget, and then just maybe they'd have a viable project on their hands.

"Any comments, Dave?"

The PD of infrastructure hadn't said much all morning. Shortly after arriving on the project, Ted had fired Dave's boss, Grant McLaren, and made Dave his new deputy director. Ted needed everyone on board, so he didn't let anyone off the hook. He knew his colleague had something to say; sometimes he just needed a push. Dave sat up a little taller and cleared his throat.

"I do think the smelter idea has some serious merit. It makes a whole lot of sense, with energy costs rising everywhere; we could save big on utilities and construction costs by tapping China's cheaper grid and competitive labor costs." Dave paused and wet his throat with a sip of water. He was one of the best systems and structural engineers Ted knew in the industry, a brilliant introvert and not the kind of guy to pull a one-liner out of his pocket protector at a cocktail party.

When it came to making the big decisions, more than anything, it mattered who was in the room. Ted had wanted Dave, not Grant. At the outset, the company's CEO, Marius Kloppers, lacked faith in Ted's call on Dave, but Ted stuck to his guns and was glad he had. Dave had a knack for cutting to the chase. He helped steer the debate in the right direction. Dave spoke his next words carefully.

"Safety and quality are manageable over in China if we have our own crew on-site. It's getting the right folks on board over here that we need to concern ourselves with. HR and legal—do we have their vote?"

Dave glanced right, down the oblong table at the external relations deputy, Richard Yeeles, who adjusted his collar and then swept a crumb off his notepad. To his left, sat a couple of his legal managers. It was their business to make sure corporate and government relations flowed smoothly. The Department of Mining and Petroleum, of course, had vested interests in keeping Olympic Dam operations in South Australia. Talk of moving the smelter to China would really get their knickers in a knot. Over the next decade, the OD expansion was expected to put $250 million into government coffers and boost South Australia's GDP by 6 percent.

"Moving a major ore-production facility overseas and outsourcing labor would seriously scale back on government revenues, not to mention domestic jobs and product, as the world sinks deeper into a recession." Richard laid out the issues even-handedly. He could handle himself with this crowd. The issues were nothing personal. His job, at this point, was to sit on the fence and play devil's advocate to make sure government interests were weighed in sufficiently.

"A decision to build the smelter in China would stand to potentially affect every South Australian right down the line, from domestic trade workers and suppliers to average taxpayers and families," he explained.

For well over a year, anticipation of the OD expansion had been fueling domestic spending and infrastructural development. Since the mid-1980s mining boom, the South Aussies' entire economy had been tied to its natural resources. Decades of hefty investment and royalty returns on domestic mining had cleared their deficit, set them on a growth trajectory, and built up their expectations and dependency on future mining developments.

"Once the ball is rolling, the momentum and combined impact of a sudden shift could wreak havoc on the South Australian economy." Richard was a soccer fanatic and couldn't resist a good game metaphor. "We need to carefully consider to what degree we want to risk a penalty by alienating a major client. They may want to ensure their hundreds of millions, but BHP wants its hundreds of billions!"

At that point, it hit home for Ted that the politics and economics of the China smelter debate were clearly at odds. Could this be a no-win, damned-if-they-do, damned-if-they-don't situation?

Dave cut in. "Hold on, there's another way to look at this, folks. True, moving a major production facility overseas would constitute a major scale-back of profit for the Aussies. But if it kept the project viable, it would save billions in otherwise lost revenue, on both sides." Dave squared his shoulders to Richard and his team. "Why wouldn't the Aussies be on board to help save the project? After all, it's their own future at stake here. If we lose, they lose. They want Olympic Dam as bad as we do!"

For the next hour, external relations hashed out their concerns. Ted sat back, fingertips lightly touching, and soaked in the discussion. His team was on top of it, but he needed them to come together. Costs were critical. Time was ticking.

"If the Aussies stick to their guns and ignore the big picture, they could break this project." Brink leaned in.

"And if BHP can't afford the price tag of a fully domestic operation, the project will likely be shelved anyway," Marco added.

At least they understood the issues. They had clarity. But soon they would have to make a decision. Success was a measure of how swiftly they came to a decision and then acted on it. John Somers had taught Ted that. Digging OD into debt was not an option. The project had already spent hundreds of millions on initial studies, which was nothing for a company like BHP. But once they were committed and moving forward, they'd have billions on the line. Change was good, until it was bad.

Ted set his mark on China. It was their only real option left. He felt it in his bones.

> **LESSONS LEARNED**
> The whole is greater than the sum of the parts. Teamwork and open communication are key to finding creative solutions.

CHAPTER TWENTY-THREE

Good Project, Bad Timing

They'd hashed out the issues. Now it was time. On day three they put the smelter idea to a vote. Before their show of hands, Ted looked directly at each of his deputies.

He spoke calmly and unequivocally. "I believe this is the only solution to the high capital cost of the project."

Sitting tall at the head of the table, Ted urged his team to choose the project over the politics and asked point-blank for their backing. Everything else was a Band-Aid solution, not to mention a precious waste of time, talent, and money.

"Let's not worry about what the state government would say about not having a domestic smelter. Given the facts, our only real choice is to step outside the box, look at the big picture, and agree to convince the Aussies to support a radical and innovative move that will benefit everyone at the end of the day."

Then Ted placed both hands purposefully on the table. "Let's do this, folks."

He stood up and his voice carried around the room. "And remember that everyone has a vote—but this is not a democracy. Raise your hand, if you support the move to build a smelter in China."

Ted lifted his arm and counted. One, two, three. Four hands. Ted frowned.

"All those opposed." Richard's arm shot up. A rush of adrenaline shot through Ted's body. For the first time, in his long career as a project director, he'd been unable to bring his group to consensus. He was forced to exercise his veto power.

"Overruled."

He'd had no choice. Ted felt a dull sense of disappointment and alienation from his group. But he was certain of his course. There was only one thing left to do. And he had to do it. The sun slipped behind the clouds and the room fell into shade. He reached for his cell and dialed his boss in Melbourne as the others watched in silence. Ted looked almost serene as he waited for Graeme to pick up.

"Hi, it's Ted. I've got some news for you."

"OK, let's hear it."

"We're going to build the smelter in China."

"Pardon?"

"We're going to build the smelter in China."

"Ah, no. I don't think you are."

"Yeah. We have to. It's the only way we can make this thing work."

"You can't, Ted."

"What do you mean we can't? We gotta do it. There's no other way."

"The South Australian government will never let you do it."

"Well, let's go talk to them. See what they say."

"They're not interested. All they care about is how to maximize the benefits to South Australia. They'll never approve it."

Graeme wasn't trying to be difficult. For someone of his high rank, he was an unusually gentle and polite individual. Right then, Ted realized that his boss wouldn't go up against the South Australian politicians. Furthermore, the two men respected one another and neither wanted to come to heads. Ted was walking a thin line, unable to turn back.

"Graeme, with all due respect, if there's any hope of keeping this project alive—and within budget—we've got to go out on a limb here, or all our preliminary work will just sit there and collect dust. This is a good plan!"

His boss started to sound impatient and increasingly tense. "Ted, it's not going to happen, not with the economy going down the tubes the way it is. The South Australians won't buy in. That's all there is to it!"

Then there was nowhere to go. Ted surrendered to the situation without giving in to its inanity. "Well, then, you need to go talk to Marius before we continue studying Olympic Dam any further."

"Excuse me?"

"Graeme, there's nowhere to go from here. Nowhere! It's a fifteen- to twenty-billion-dollar project—if not higher. I'm telling you—if we do up a reestimate with all operations based in South Australia, the CEO will take one look at it and shelve the project!"

The uncharacteristic bitter edge in Ted's voice caught his boss off guard. Graeme wasn't a wade-into-the-details kind of guy.

"Ted, I'm sorry. No. That's final."

So the call ended, before any real damage could occur. The two men respected each other, after all. Business was business. Afterward, there was nothing left to say. Ted released his team, and when they'd all filed out, he drew the blinds and sat alone for a while in the darkened boardroom.

There were billions, if not a trillion-plus dollars of proven resources in the ground at Olympic Dam. One way or another, Australia would mine it. And, when the fiscal tides turned, they'd be right back where they started, if not deeper in the hole. Then they'd likely just have to repeat this little cost-savings exercise, hoping for different results. Wasn't that the definition of insanity, according to Einstein? Ted sighed and shook his head. Then he stood up to leave. He'd done everything he could, but his best wasn't enough.

* * *

"So what happened?" Jim asked from behind the newspaper.

Ted took a sip of cool lemonade then set down his glass. "You wouldn't believe it." He shook his head and then ran a hand through his hair.

Jim rustled the pages of the business section.

"It's right here. I just read something about Olympic Dam." Jim scanned the previous page. An article in the newspaper that morning reported that BHP Billiton had pulled the plug yet again, after spending hundreds of millions more on the OD expansion. After the explosion at Japan's Fukushima Plant during the March 2011 tsunami, the future of nuclear was unclear.

"That's right. The uranium market dropped out after that, and so they shelved the project again. Same old story. A cost-savings mandate until they could tell where things were going."

Ted and Jim were sitting on the Cooke's shady veranda, on the second-to-last day of their vacation together. The fans were spinning at full throttle. It was another scorcher. Later that afternoon, they were heading over to Quail's Gate Winery, where they'd be dining in the vines. The ladies were inside getting ready.

Ted had since come to terms with the outcome of the Olympic Dam project. However, had he made one decision differently that final day, they could have possibly had a different outcome.

"My boss never did run the idea of a China smelter past the CEO, but Marius eventually caught wind of it and..." Ted leaned in.

"And?"

"You know what he said?" Ted looked at Jim intently from under the tops of his bifocals. "He said, 'That's the best damn idea I've heard in years!'"

"Unbelievable." Jim sat back in his seat, crossing his arms. He considered this a moment. "If only you'd gone above your boss's head."

Ted nodded. "I wish I had, for the sake of the project."

He might have offended Graeme, but if he'd been willing to break the chain of command and pushed the issue, they might have gotten their China smelter after all.

"So what did this Marius fella do?" Jim wondered.

Ted filled in the details. "When he learned about the smelter idea, he actually jumped on board with it. He took it up briefly with South Australia's premier, Mike Rann. He called it 'the only economic option they had.' I knew we were onto something, but by that point it was too late."

The idea fizzled at the corporate management level, partly because the project had lost serious time, and also credibility issues had emerged between the project team and head office. In the end, Ted's moral compass took him one way, when they probably needed to go another way.

"I think partly to blame was the whole corporate culture at BHP."

"You mean penny-pinching?"

"Yeah. They were still running things like a small-time mining outfit that had cash-flow problems. In reality, the company just hadn't grown into its big shoes yet."

Jim nodded. "What ever happened to that fellow, Jake, who made you spit the dummy, or whatever you call it?

"Oh, Jake." Ted shook his head.

His first regret on OD was not taking the smelter idea to Marius. His second was how he'd treated Jake. In one way, his method had proven effective, but in another he'd undermined one of his own management principles: when dealing with team members, criticize in private and compliment in public.

"I removed him from the team. I felt a little bad about that, but you know, if you've got the best technical person in the world, but he's not a team player, I don't want to work with him."

"Did you let him go?"

"No, I had him placed back into a high-level technical job much more suited to his temperament and skill set." An exceptional tradesman, yes, but a senior deputy director, Jake was not.

"And what about you, did you leave the project?"

"Eventually." Ted paused, taking a sip of his drink. "After Olympic Dam was shelved, my boss's job became redundant so he was let go. I stayed on another year, continuing to study ways to reduce capital costs and work on the environmental impact report, which was needed for government approvals."

"Why did you bother sticking around?"

"I wasn't happy with how things turned out, but I wasn't going to break contract and jump ship."

"Makes sense." Jim nodded. "You've got to finish what you start."

Ted sighed. "I tried."

But sometimes what looks like a great project, just isn't. Jim looked at his friend, at first puzzled but then with empathy in his eyes.

"I did what I could, and then Anita's health took a turn for the worse, and Mona and I moved home permanently."

> **LESSONS LEARNED**
> It is in the best interest of the corporation to have accurate information on the cost and schedules of projects. Accurate cost and schedules are based on having a valid defined scope.

CHAPTER TWENTY-FOUR

October 5

Two events in Ted Bassett's life profoundly changed him, and they uncannily occurred on the same day fifteen years apart.

The Soviet Mi-8 chopper carrying fifteen passengers and crew crossed the Barskoon Pass, headed toward Bishkek through the rugged and winding Tien Shan Mountains. The sky was partly cloudy.

The leaves were turning color the fall day Anita checked into the hospital with severe pain in her back and shortness of breath. Her life-extending drug had stopped working.

Once inside the mountain range, the helicopter encountered fog and flash lightning. The local Kyrgyz pilot and his crew tried to find their way out of the treacherous mountain maze with little else at their disposal but a stopwatch and working altimeter.

Upon her admission, the doctors checked Anita's blood cell markers and heart rate and did their routine scans. By this point, the cancer had spread too far. They told Anita that she had little time left and to go home. The fight was over. She was dying.

Finding themselves in unexpected and life-threatening flying conditions, the pilot and crew aboard the Mi-8 did what they could to make it safely to the other side. Without warning, the helicopter crashed into the mountainside, instantly killing all onboard.

With her family by her side, Anita took her last breath.

CHAPTER TWENTY-FIVE

Friends

"To us!" Beryl raised her glass joyfully. "To an absolutely wonderful week together!"

It was a perfect summer day. Hot, blue sky, gentle breeze. Beryl looked with affection to Mona and Ted and then to Jim, who all followed suit, extending their arms. As they clinked their fine crystal stemware together, a sweet-sounding harmony rung out from the balcony of Quail's Gate Winery.

"A good time had by all." Jim nodded his head.

"So relaxing," Mona said with a happy sigh. "At least for us, girls, anyway."

She exchanged playful glances with Beryl, whose hand flew to her hip. "Yeah, all you guys did was sit around talking about work. How fun is that?"

Jim smiled sheepishly, and then winked knowingly at his cousin-in-law, who burst out laughing.

"Actually, we really enjoyed ourselves. We got a lot...accomplished."

Beryl rolled her eyes. "Well, that's good, at least. We did too." She and Mona were each sporting new outfits from their power-shopping excursion at the outlet mall that afternoon. They linked arms for a moment. Mostly they'd just loved having the time to catch up with one another.

Jim whistled, and Ted chuckled. "You lassies look great." Then he turned to look out over the rolling green vineyards descending toward Okanagan Lake.

Friends

"Isn't this just beautiful?"

"Stunning." Mona moved to lean against the railing.

Live jazz music drifted up from the patio where the open-air festival was shifting into full swing. She swayed her hips and a smile curled her lips. In a couple of hours, they'd be treated along with eighty other guests to an exclusive five-course meal amid the grapevines as the sun set over the lake.

"This is one of our favorite places on earth," Jim said, matter-of-factly, as he scanned the landscape with the kind of appreciation an engineer has for a smooth-running operation. Even he knew deep down that all the work in the world mattered little compared to the simple pleasures in life, such as times like these, and family.

"We try to come at least two or three times a summer," Beryl said, "whenever we have friends in from out of town. The setting's just beautiful, and the food here is exceptional." She kissed the tips of her fingers. Quail's Gate was considered to be one of the best dining experiences in the Okanagan.

Mona turned toward Beryl. "Hey, didn't Anita and Mike join you here a few summers ago?" She and Ted had been living in Adelaide when their daughter and son-in-law had taken a couple weeks to do a summer bike tour of British Columbia, stopping in the Okanagan to visit a few wineries.

"That's right. They stayed a couple of days with us, and we came here. Come to think of it, Jim, wasn't it for this exact jazz festival?" Beryl turned to her husband.

"Yes, I think so. You're right. We did the festival, not the dinner, but we spent the afternoon down on the lawn, right over there." He pointed across the yard toward the bandstand.

Beryl's voice grew excited. "Yes, that's right. Anita loved it! She and Mike danced up by the stage all afternoon. We had fresh cherries and peaches and champagne."

Mona's eyes brightened. Her hand flew to her throat. She could picture her daughter right there, soaking up the same scene they were enjoying that very moment, a giant smile on her face, glistening like the sun's sparkle dancing on Okanagan Lake. She looked at Ted, who, just for a split second, appeared as if he'd seen a ghost. But then his face cleared, and a deep peace fell over him. The gift of a new memory for a parent who'd lost a child was precious beyond any words. Then all he heard was the music. Ted smiled broadly, and an enormous weight lifted from his shoulders as he felt his daughter's presence around them.

A few seconds delay, and then, he finally spoke. "Wow! Really? What were the chances?"

Out of nowhere, an idea suddenly struck him. He turned to his wife, catching her by surprise. "Mona, remember how Anita always said she wanted me to write a book?"

"Yes, of course! About all of your crazy, far-flung adventures in project management?" She looked amused, but then realized her husband was completely serious.

Jim's ears perked up. "You should do it."

Ted turned to his cousin-in-law. "You should help me."

Jim cocked his head, seriously considering. He scratched his bald patch as he spoke: "The industry could use it that's for sure—with all the senior management people retiring and the new folks coming in wet behind the ears."

As he spoke, he twisted the beveled engineering ring on his finger, and Jim's mind started churning out the first chapter. "You could call it the Principles of Project Management—or something like that." Beryl held her tongue.

"I like it, I like it. Keep talking," said Ted.

"Another project?" Mona looked at her husband. "You can't be serious?!"

"Why not? It could be a retirement project." Ted grinned, growing even more excited at the idea. Right in that moment, he knew he had to do it. Long ago he'd promised Anita that he would. Success could build you up and serve to inspire greatness in others. But it was the extreme challenges and failures that broke you down, that tested your character, your strength and resolve. For Ted, that's where the real learning began, where dreams faded and reality set in. Then projects and possibilities grew spontaneously out of the great potential that lay within. Anita had taught him that.

"But you guys aren't even retired," Beryl protested, halfheartedly.

"It could be a little something on the side." Jim eyes danced, and Beryl slapped her husband's behind. They still had it after all those years.

Both men shrugged, and then looked each other in the eyes, mischievously.

"Let's do it."

Jim raised his wineglass, determination in his voice. "To the book."

Ted smiled triumphantly and raised his glass. The sun refracted through the vessel, turning into dancing colors that touched him that summer day just where he needed it the most.

"To the book. For Anita."

APPENDIX I

Project Management Principles

Before Starting the Project
1. Engage the strongest project leader possible. A great project leader can save millions on a project. This being the case, the leader's salary is incidental to the cost of the project. Strengths should include relevant experience with the industry and size of project. The leader must demonstrate leadership and team motivation. A micromanagement manager never works on major projects.
2. Let the project leader hire the best people possible and challenge them by delegating the maximum authority possible.
3. Salary should not be an overriding factor.
4. Once you have the right team, keep it together. Changes at the senior level are a major reason for project failure.
5. Develop an overall master program that summarizes all project activities in an easily readable format.
6. Develop an organization chart with clear lines of authority.
7. Do not authorize significant capital funds until the scope, schedule, and estimate are agreed. Starting capital projects with poor definition is a major cause of project failures.
8. Develop contingency plans in case of cost overruns or schedule delays.
9. Develop key performance indicators for the team based on achieving project milestones.

Once the Project Is Underway

1. The impact of changes must be carefully evaluated. Change should not be implemented unless it has a significant positive impact on safety, schedule, and/or cost.
2. Make timely decisions. Procrastination stalls progress.
3. Maintain a balance between gathering additional detailed information (analyses can cause paralysis) and decision-making.
4. Establish good, open communications and reporting. Keep senior management not directly involved in the project fully informed. Do not hide problems hoping they will disappear; they will only get worse.
5. Hire specialists as required. Do not try and reinvent the wheel.
6. Never have more than one technical innovation on a project.
7. Remove nonperformers quickly or a cancer will spread within the team.
8. Do not consensus manage. Everyone should have a say but the project leader has more votes than the rest of the team put together.
9. Do not get caught up in the belief that project systems will save the project. Good people will make a bad system work, but a good system will not make poor people perform.
10. Be commercially fair to subcontractors and suppliers. Make them part of the team. This does not mean "alliance" or "partnership" or "no hard money" contracts.
11. Avoid joint ventures. Difficulty in making decisions goes up exponentially with an increase in the number of companies involved in decision-making.
12. Don't let the bureaucracy grind the project down. Presentations to management that require decisions should be kept to one page with detail backup supplied only if requested.
13. Managers must be directly involved in developing project plans. *Don't give the project planning to schedulers.* They can develop the details once the formal plans are made.
14. Standardize where possible.
15. When dealing with team members, *criticize in private and compliment in public.*
16. Celebrate victories and analyze failures for learning.
17. Use leading indicators to analyze project status. Report history.

Portfolio of Pictures

Portfolio of Pictures

Beach at Adelaide

Construction at Diavik

HOW TO BUILD A PROJECT

Diavik Setting Sun

Diavik Construction Camp

Portfolio of Pictures

Construction at Kumtor

Kumtor Executive

HOW TO BUILD A PROJECT

Drift through Glacier at Kumtor

Kumtor Management Team

Portfolio of Pictures

Kumtor Site

Road to Kumtor

HOW TO BUILD A PROJECT

Sampling the Kumtor Ore

Summit - Road to Kumtor

Portfolio of Pictures

Kyrgyz Helicopter

Project Director at Work

HOW TO BUILD A PROJECT

Original Construction Office

Ted Bassett and John Somers

About the Authors

TED BASSETT graduated from the University of Saskatchewan in 1968 with a bachelor's degree in mechanical engineering and in 1970 with a master of science. He has forty-one years of experience in the mining and metals business: fifteen years operating companies and twenty-five years in engineering contracting. He is currently operating a small consulting company. Ted sits on the board of directors of Noront Resources, Ltd., a junior mining company.

JIM COOKE graduated from the University of Saskatchewan in 1959 with a BSCME, whereupon he joined Shell Canada. He started his career in oil and gas exploration and production and at an early stage was transferred into a subsea research group within Shell US. This led to a career in offshore projects at home and overseas. Throughout his career, he chose to be an engineer's engineer for the development of technical people and their ability to think and function within large offshore projects where both the risks and the rewards are high. Jim lives in Kelowna, British Columbia, with his wife, Beryl, surrounded by a wonderful family of thirteen.

JENNIFER PARKS is a freelance journalist and creative writer. She is the author of *Canada's Arctic Sovereignty: Resources, Climate and Conflict* (Lone Pine Publishing, 2010). A former Sun Media reporter and Honours Cultural Studies graduate of McGill University, Jennifer lives in Kelowna, British Columbia, with her husband and two cats.

Made in the USA
Charleston, SC
15 October 2015